HUMAN FACTORS/ ERGONOMICS FOR BUILDING AND CONSTRUCTION

CONSTRUCTION MANAGEMENT
AND ENGINEERING
Edited by John F. Peel Brahtz

HUMAN FACTORS/ ERGONOMICS FOR BUILDING AND CONSTRUCTION

Edited by
MARTIN HELANDER
Canyon Research Group, Inc.

A Wiley-Interscience Publication
JOHN WILEY & SONS,   New York · Chichester · Brisbane · Toronto

Library of Congress Cataloging in Publication Data:

Main entry under title:
Human factors/ergonomics for building and construction.
  (Construction management and engineering)
  "A Wiley-Interscience publication."
  Includes bibliographical references and index.
  1. Human engineering.   2. Engineering—Manage-
ment.   3. Construction industry—Management.
I.Helander, Martin, 1943–

TA166.H79   624   80-26717
ISBN 0-471-05075-X

Printed in the United States of America

10 9 8 7 6 5 4 3 2 1

SERIES PREFACE

Industry observers agree that most construction practitioners do not fully exploit the state of the art. We concur in this general observation. Further, we have acted by directing this series of works on Construction Management and Engineering to the continuing education and reference needs of today's practitioners.

Our design is inspired by the burgeoning technologies of systems engineering, modern management, information systems, and industrial engineering. We believe that the latest developments in these areas will serve to close the state of the art gap if they are astutely considered by management and knowledgeably applied in operations with personnel, equipment, and materials.

When considering the pressures and constraints of the world economic environment, we recognize an increasing trend toward large-scale operations and greater complexity in the construction product. To improve productivity and maintain acceptable performance standards, today's construction practitioner must broaden his concept of innovation and seek to achieve excellence through knowledgeable utilization of the resources. Therefore our focus is on skills and disciplines that support productivity, quality, and optimization in all aspects of the total facility acquisition process and at all levels of the management hierarchy.

We distinctly believe our perspective to be aligned with current trends and changes that portend the future of the construction industry. The books in this series should serve particularly well as textbooks at the graduate and senior undergraduate levels in a university construction curriculum or continuing education program.

JOHN F. PEEL BRAHTZ

La Jolla, California
February 1977

v

PREFACE

This book is the first of its kind in the construction industry. It symbolizes the growing interaction between social scientists and engineers. The objective is to present important information taken from the social sciences, in particular human factors and ergonomics, and apply it to construction work.

The text covers a broad range of topics such as: organization of construction work, measurement of productivity, work satisfaction, construction site safety, health risks at a construction site, effects of heat or cold on comfort and productivity, ergonomic design of construction machines, use of training programs, and the effects of affirmative action and minority rights.

The book is primarily written for professionals such as civil engineers, managers, and safety personnel in the building/construction industry. It is also recommended as a textbook for senior undergraduate and graduate levels in a university curriculum, or for use in continuing education programs.

Much of the information presented is very general in nature and has applications outside the construction field. This is not surprising since the book deals with the capabilities, limitations, and vulnerabilities of human beings. The text is thus well suited not only for construction engineering students, but also for systems engineers and students of human factors and industrial psychology.

The construction industry is the largest single industry in the United States, supplying about 10% of the Gross National Product. Unlike other industries, it is predominately comprised of small, local companies which lack the financial resources to support research and development. This certainly holds true for research in human factors, productivity, and work satisfaction. Many of the theories presented therefore rely on findings from research in other industries, mainly the manufacturing industry.

This book is not only the first of its kind in the construction industry but is also the first in the human factors field aimed at one particular professional group. A broad interdisciplinary coverage was made possible by inviting distinguished specialists to present their area of expertise in a concise and easy-to-read format. I am deeply grateful for the contributions to this text supplied by Drs. Englund, Horvath, Levitt, Miller, Safilios-Rothschild, Söderberg, Taylor, and Zenz.

The book would not have been written were it not for the advice of Dr. John F. Peel Brahtz, editor of the Wiley-Interscience Construction Series. I owe him my deepest gratitude for encouragement and for patience in waiting for the complete product.

I began organizing this book while working at Human Factors Research, Inc. in Santa Barbara and completed it while working for Canyon Research Group, Inc. in Westlake Village, California. It is a pleasure to acknowledge the formal support that both companies provided me, and in particular I want to thank Dr. Robert Mackie and Mr. John Merritt for their valuable guidance concerning the organization of the text.

Several persons have provided major assistance in typing and editing the draft manuscripts. Lynda Lee Chilton was particularly helpful in starting the project. I treasure her keen sense of organization and her unique typing skills. Salena Kerr contributed a powerful combination of writing talent and editorial logic that made her an extraordinary asset in this endeavor. Jayne Schurick demonstrated never-failing patience in typing and providing editorial assistance. I am deeply indebted to them all.

Finally I want to acknowledge the Swedish Work Environment Fund, which provided financial support for the writing of Chapter 2.

MARTIN HELANDER

Santa Barbara, California
March 1981

CONTENTS

CHAPTER 3   PHYSICAL HEALTH HAZARDS IN CONSTRUCTION   53

CHAPTER 4   CHEMICAL HEALTH HAZARDS IN CONSTRUCTION 81

CHAPTER 5   WORK PHYSIOLOGY 109

CHAPTER 6   HUMAN FACTORS ENGINEERING IN CONSTRUCTION WORK   141

CHAPTER 7   THE PSYCHOLOGY OF JOB
SATISFACTION AND
WORKER PRODUCTIVITY 183

HUMAN FACTORS/ ERGONOMICS FOR BUILDING AND CONSTRUCTION

CHAPTER 1

INTRODUCTION

MARTIN HELANDER, Ph.D.

Canyon Research Group Inc.
Westlake Village, California

The objective of this book is to sort out important information from the social sciences and present it so that it can be easily understood and applied to construction work.

1.1 BACKGROUND

In most countries around the world, construction is the largest single industry both in economic output and in the number of people employed. In the United States the construction industry directly employs about 5% of the total labor force and provides about 10% of the gross national product (GNP). Considering the importance of the construction industry and the available economic resources, it is surprising that there has been so little research and development on the human side of construction work. The intent of this book is to stimulate interest in this very important area. We will try to provide answers to questions such as the following:

- How should a construction company be organized?
- Is it possible to measure construction worker productivity?
- What can be done to organize a crew so that each individual is satisfied with his work?
- How is the individual worker encouraged to be productive?
- What can be done to ensure construction site safety?

- How do accidents affect productivity?
- What are the important health risks at a construction site?
- What can be done to prevent cancer, loss of hearing, or loss of eyesight?
- How much physical work can and should an individual perform?
- How does working in heat or cold affect productivity and comfort?
- How should a construction machine be designed so that it is easy to operate?
- How does shift work affect productivity and worker comfort?
- Are there any reasons why a construction company would use training programs?
- Does affirmative action affect construction productivity?

All of the above are concerns of human factors specialists. Human factors engineering is a fairly new science. It has its roots in the time and motion studies that originated during the World War I. Gilbreth, the main proponent of this science, was originally a contractor. In 1909, he successfully pioneered the application of time and motion studies to masonry and found that mason productivity could be increased by setting size and weight standards for bricks and by limiting the height that they should be lifted. Since Gilbreth's studies, there has surprisingly and unfortunately not been much further research in human factors applied to construction work.

Human factors engineering, however, has prospered as an independent, interdisciplinary science ever since World War I. People working in human factors accumulated knowledge and theories from many other disciplines, such as experimental psychology, industrial psychology, work physiology, and the systems sciences. Any piece of knowledge or theory from the social sciences that had the potential of application in the real world has eagerly been absorbed. Unfortunately, most of the research in the social sciences has been "basic." This tradition is in many academic circles preferred to applied research, to the regret of most engineers.

During and after World War II, there was much interest in military applications of human factors. In fact, military resources have since provided the majority of research funding (maybe 65%) in the human factors field. The military interest was inspired by the realization that several of the newly developed war machines were inhuman in ways other than the obvious. They were often so complex in their design that human operators could not handle them (e.g., controls and displays in airplanes and tanks) or they introduced tasks that were so boring the operators often fell asleep after 30 min of work (e.g., radar or sonar surveillance).

Much of the research findings from these military applications can fortunately be translated directly to other industries; the operator remains the same, although the environment changes.

During the 1960s and 1970s, there has been a considerable amount of human factors research in traffic safety. This research has addressed problems of design for both vehicles and roads. It has also investigated areas such as crash worthiness of vehicles, driver training, and effects of drugs. In highway design, for example, there are two human factors principles that are consistently used:

1. The implications of drivers' reaction times for designing horizontal and vertical curves.
2. The use of a transition curve to smoothe the transition from a straight road to a curve of a specific radius.

The difficulty in applying social science research is the main reason it has generally been rejected in engineering school curricula. The social sciences generally do not teach practical applications and the engineering sciences do not think an engineer should be taught "soft" subjects (Larew, 1978).

The only significant input to human factors in construction has come from architects. Even so, they are generally more interested in the design aspects of buildings than in the human factors considerations in constructing them. Very little research has been done to investigate how working conditions can be improved. This seems even more striking in view of the comparatively larger efforts made in the manufacturing industries, such as studies of assembly line organization, the effects of fatigue and shift work, design of inspection tasks, and ways of achieving worker satisfaction, work motivation, and productivity. All we know from research in industrial environments has been initiated by the manufacturing industry.

Why, then, has the construction industry not responded with the same interest as the manufacturing industry? There are several reasons:

1. In general, the construction industry does not spend resources on research and development in any area. Even the major, most important innovations, such as building cranes, modern drilling and blasting techniques, and administrative planning systems, have come from supporting industries rather than directly from the construction industry. Only 0.3% of the annual volume is spent on R and D whereas 10 times

as much (2.2%) is spent by the manufacturing industry (Tucker and Borcherding, 1977). One reason for this is that most construction companies are small in size and understandably do not have resources for independent research and development. (Companies with less than 50 employees employ about 65% of the work force.)

2. Construction work is generally regarded as a temporary commitment. This is certainly true with respect to a particular job. (In addition, construction workers have a turnover rate that is slightly higher than that of other industrial workers.) Employers therefore do not feel it is important in the long run to assume short-term responsibilities for safety and health, working conditions, and machine control design, for example.

3. Construction workers have a higher degree of work satisfaction than many other industrial workers. This is due to the fast feedback that building work provides. In addition, the trade has a long tradition of working in small groups with autonomous decision making. This basic satisfaction reduces workers' tendencies to complain about human factors types of problems.

4. A slightly more speculative reason is that the construction trade itself is fairly conservative. This might reduce the likelihood of adopting novel interdisciplinary methodologies.

The temporary nature of most building jobs and the basic satisfaction with work reduce the tendency of workers to complain about existing work hazards. This might have serious consequences since health disorders can develop over a period of years (and across several work sites) without previous notice. Examples of these long-term effects are the development of skin diseases from exposure to cement, low back pain from incorrect lifting postures, or loss of hearing owing to overexposure to noise. All of these are cumulative health problems, and workers can be exposed for years unless someone at a building site has the knowledge and the authority to make the necessary corrections. This ideally would be the worker himself.

A number of countries in Europe (for instance, West Germany and Sweden) are presently providing federally funded training to workers so that they can "look out for themselves." In these countries, it is considered an equal rights issue to be informed about the hazards of work and to initiate necessary countermeasures.

Since there has been so little research it is difficult to estimate the magnitude of most human factors problems in economic terms. A couple of studies, however, provide estimates of the cost of construction accidents. Levitt (1975) estimated that in 1972 the costs of construction accidents, conservatively estimated, amounted to $8.2 billion. Of this sum, only $1.6 billion was reimbursed by insurance companies. Helander (1980) estimated that the cost of construction accidents might be close to 1% of GNP (if Congress only knew!).

Accidents on the construction sites are a problem of very great economic interest to a construction company since (if nothing else) they increase workers' compensation premiums. A few construction firms in California have recognized this fact and are promoting safety on an equal basis with costs and productivity. This has led to a dramatic reduction in accidents, a finding that substantiates the importance of company attitudes. As a result, there has been a decrease in the workers' compensation premiums that is of the same magnitude as the companies' profit margin (if managers only knew!).

1.2 ORGANIZATION OF THIS BOOK

This book is intended to provide information about how human factors engineering can be applied to construction work. The book is written primarily for construction managers and civil engineers, but it should serve well as a textbook in a university curriculum or continuing education program.

Each chapter introduces a minimum of background material of a theoretical nature. This information is regarded as necessary to fully comprehend the applications and how they can and should be made.

There are three major sections: environmental hazards, operator factors, and organizational factors (see Figure 1.1). Environmental hazards are dealt with in Chapters 2, 3, and 4; operator factors in Chapters 5, 6, and 7; and organizational factors in Chapters 8, 9, and 10.

Chapter 2—Safety in Construction

This chapter starts by analyzing causes of construction accidents. Usually an accident occurs because several (unexpected) events happen at the same time. Although worker "error" is usually the precipitating cause of an accident, there is a need to make the environment safer. Rules for how this can be achieved are mandated by the OSHA guidelines, the most important of which are summarized.

Construction work is one of the most hazardous occupations in the United States. From accident statistics it is concluded that the most common type of accident is falling of a person or materials.

The best way to prevent accidents is to change company attitudes

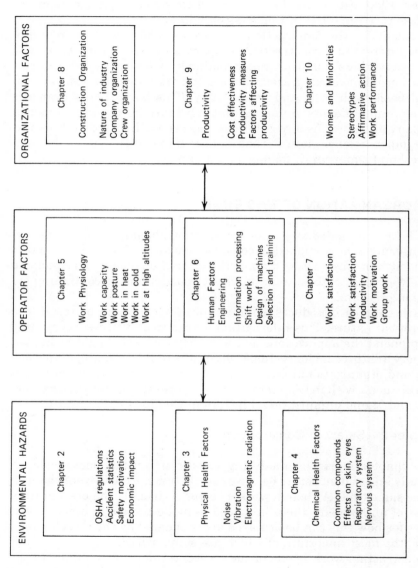

Figure 1.1 Organization of this book.

toward accidents. Accidents are really crucial both from humanitarian and economic points of view. Once management realizes this and takes proper action, for example, by charging accident costs to individual projects, there will most likely be a drastic reduction in the number of accidents.

Chapter 3—Physical Health Hazards in Construction

This chapter deals with the effects of exposure to noise, vibration, and radiant energy. Perhaps because it is so obvious, exposure to noise is often considered one of the major work hazards. Noise is irritating and makes it difficult to talk, listen, and even think. More important, however, is loss of hearing, which develops gradually for persons exposed to noise over a long period of time. Medical checkups can be used to identify people who are particularly subject to hearing loss owing to noise exposure. Other precautions include the use of ear protection devices that effectively reduce the noise and the installation of sound absorbing insulation at the work site.

There are two types of vibration, whole body and segmental. Whole body vibration, especially in the region of 3 to 5 Hz, is perceived as uncomfortable. Exposure to whole body vibration can lead to indigestion, low back problems, and an increase in blood pressure. Light persons usually have less trouble than heavy persons.

Hand-held tools such as concrete vibrators or pneumatic devices induce segmental vibration of the hands. This can lead to circulation problems manifested as Reynaud's disease and Dart's disease.

Electromagnetic radiation can lead to several medical disorders. This chapter discusses health effects and exposure limits (threshold values) for infrared radiation, ultraviolet radiation, microwaves, and lasers.

Chapter 4—Chemical Health Hazards

Hazardous chemicals may enter the body in one of three ways: by inhalation, by ingestion, or by absorption through the skin. This chapter explains how chemicals affect the body organs and how they can cause medical disorders such as allergies, skin diseases, and, in the worst case, cancer. There are several compounds that are potentially dangerous if handled improperly: form oil, concrete, mineral wool, asphalt, caulking agents, adhesives, and plasters.

There are two ways that the hazardous effects of these chemicals can be investigated: through laboratory observation of the effects on animals and through analysis of health statistics. Most of the existing knowledge comes from laboratory research and studies of health conditions among workers who manufacture the hazardous substances. Little has been concluded from direct observation of health records of construction workers. This is due to the difficulty in obtaining data on exposure times.

Chapter 5—Work Physiology

Construction work involves a great deal of physical activity. This chapter provides basic information on the functioning of the human body as it performs work.

Work capacity depends on factors such as conditioning, familiarity with the task, sex, and age. Most of the tasks in construction work involve moderate levels of physical work and can easily be performed by women and older workers. Some of the environmental factors that may affect productivity by providing stress on the worker are discussed in this chapter.

Working in a hot climate can be very taxing and results in lower productivity. In a very hot and humid climate it is therefore necessary to reduce the work load. A method for measuring heat stress is presented.

Similarly, working at a high altitude induces several physiological changes and reduces work capacity. If compensatory actions are not taken, mountain sickness or pulmonary edema may result.

Chapter 6—Human Factors Engineering

Some key human factors areas are presented in this chapter: shift work, illumination, design of machines, and selection and training.

Shift work is becoming increasingly common among construction workers. Most workers adjust to night work after 1 week. However, the social effects of shift work are difficult or impossible to ameliorate. The effect of different shift work schedules on the body rhythms and social factors is discussed.

Construction sites are often poorly illuminated. Older workers especially may have problems in seeing important details of the work. Increasing illumination levels is a cost-effective measure that promotes job satisfaction as well as productivity.

Construction managers are often in the position of ordering new machinery. The performance aspects of the machines naturally play an important role in the decision making. The "man–machine interface" is often equally important but difficult to evaluate unless the decision maker knows both the worker and machine capabilities. This section provides some basic information, quoting relevant SAE standards for construction machines.

A common practice in many construction companies is hiring and firing workers until the "right" mix of skills is obtained (often by default). This practice is unsatisfactory for both the workers and management. Techniques for the selection and training of workers are available. These represent a better alternative than current practices.

Chapter 7—The Psychology of Job Satisfaction and Worker Productivity

Several theories of work satisfaction have been proposed during the last 25 years. The best known are those developed by Maslow and Herzberg, who were among the first to study nonmonetary incentives for work. Later, Vroom described how job satisfaction results from a good match between an individuals' needs and the goals of the organization.

Management sometimes fails to realize that workers' attitudes about their jobs are usually not formed at work, but in their outside social environments, through parents, school, friends, mass media, and so on. Management efforts to boost worker motivation therefore may have little effect.

Any program to change employees' attitudes, morale, satisfaction, or productivity must take these social factors into account. Some guidelines are presented to aid a construction manager in this difficult process.

Chapter 8—The Organization of Work in Construction

The intent of this chapter is to describe the organization of construction firms, projects, and crews and to highlight some of the ways in which construction work is structured differently from work in other industries.

The organization of construction firms depends on the characteristics of the constructed facilities and the resources available for constructing them. The close relationship between design practice, labor and materials markets, and work organization results in a considerable degree of inertia against change. This may help to explain why the industry has so often been accused of failing to innovate.

The chapter begins by pointing out some of the unique aspects of the construction product such as immobility and the unique requirements for each individual job. Subsequent sections analyze how firms, projects, and crews are organized so that they can match the demands of the job to resources and manpower. We also devote considerable attention to the motivational aspects of the construction crews and an analysis of performance evaluation.

Chapter 9—Construction Productivity

Construction is a very large, fragmented, and complex industry. It is therefore hard to pinpoint problems within the industry. There is, however, a growing awareness of the importance of increasing productivity in construction. Unfortunately, current R and D efforts are minimal. This limits the productivity increase to about 1%/year. Productivity increases have largely been due to the development of more efficient tools and machines. This is a result of R and D efforts in the *manufacturing* industry.

There are several common misconceptions about the mechanisms of the construction industry which make improvements in productivity (and R and D) hard to achieve:

1. The industry is craft oriented and hence reluctant to adopt mechanized or automated building procedures.
2. Each new facility must be unique.
3. Unions will not cooperate in achieving increased productivity.
4. There is now way to obtain productivity data.
5. Nothing can be done about the liability issues that stifle the utilization of new technology.
6. Research and development apply to all industries except construction.

This chapter attempts to resolve these misconceptions and identifies ways to measure, analyze, and improve productivity.

Chapter 10—Women and Minorities in Construction: The Impact of Affirmative Action and its Effect on Work Productivity

In many countries, construction is an occupation dominated by women. This is especially true for countries where construction work is poorly paid and has low prestige. In countries such as the United States, where the occupation is formally organized and/or upgraded in terms of pay or prestige, men enter the occupation, displacing women.

The formal organization of construction has also kept minority men out of the highly skilled trades, where apprenticeship and union membership are more essential. The greatest concentration of black men has been among construction laborers who perform the heaviest and lowest paid jobs.

Legislative pressure to widen access for minorities began in 1969 but similar pressures for women did not seriously begin until the late 1970s. Recent legislation has mandated quotas for hiring female and minority construction workers.

This chapter explains common stereotypes in the United States held against women and minorities in construction work. The processes and mechanisms that have kept them out of the market are explained in detail.

REFERENCES

Gilbreth, P. B. *Bricklaying System.* New York: M. C. Clark, 1909.

Helander, M. Safety Challenges in the Construction Industry. *Journal of Occupational Accidents*, **2**, 257–263, 1980.

Larew, R. E. Construction Engineering and Management. Columbus, Ohio: The Ohio State University, Personal communication. 1978.

Levitt, R. E. *The Effect of Top Management on Safety in Construction* (Rep. 196). Palo Alto, Calif.: The Construction Institute, Stanford University, 1975.

Tucker, R. L. and Borcherding, J. D. Contractor Attitudes toward Construction R and D. *Journal of the Construction Division ASCE*, **103**, 465–478, 1977.

CHAPTER 2

SAFETY IN CONSTRUCTION

MARTIN HELANDER, Ph.D.

Canyon Research Group Inc.
Westlake Village, California

In the history of civil engineering there have been many disasters involving the collapse of structures in the process of construction. Such disasters usually result in numerous fatalities and major economic losses. They make front-page news; most of the readers of this book will recall seeing pictures of bridges or tall buildings collapsing. An accident attracts the attention of the news media if many people are killed, if a well-known person is killed, if it is bizarre or spectacular, if it is difficult to explain, or if, as with the failure of a bridge, it suggests a risk to the general public. Although the protection of the general public is important, this chapter is concerned primarily with the safety problems of construction workers, a subject that usually does not make headlines but is of great importance. To this end we review the most relevant of current safety regulations and analyze technical measures for preventing accidents at construction sites. Finally, we look at the human side of accident prevention by considering problems with attitudes and administration of accident prevention programs.

2.1 THE MAKING OF A DISASTER

"Construction Scaffold Collapses. 51 Workers Plunge to Death." This was the headline of an Associated Press news release dated April 27, 1978.

"All 51 workers atop a scaffold inside a power company cooling tower

13

were killed when the construction framework collapsed and the crew
fell 168 feet to the ground in a twisted mass of steel and rubble" (Associated Press news release, April 27, 1978).

John Peppler was standing on the ground in the middle of the tower
when the disaster occurred; he said that the scaffold wrapped around
the inside of the tower began peeling away from the wall and then fell.
"The first I heard was concrete falling. I had just sent a basket of
concrete up. I looked over my left shoulder and I could see it falling
through the air and everything falling." Peppler said he jumped under
a truck ramp inside the tower and the four other workers with him
ran to the center of the tower. All those on the ground escaped injury.

The cooling tower was being built at the plant site near the Ohio
River. The tower was to be 430 ft tall and 360 ft at the base when
completed. George Morrison, an engineer, said, "I was on the ground
when it started making noise. It was just a roar." Another witness said
of the doomed men, "They knew what was happening but there wasn't
anything they could do about it. They just fell like dominos. I looked
up and men were screaming and hollering." One worker, Leo Steel,
lost four of his five sons, a brother, two brothers-in-law, and a nephew.
The surviving son was also employed at the construction site but was
not on the scaffolding.

The Occupational Safety and Health Administration (OSHA) responded immediately to the catastrophe. OSHA officials came to the
site to determine what had caused the accident.

A side view of the cooling tower, shown in Figure 2.1, shows how
the work was performed. Construction of this cooling tower was being
done with a gliding type of formwork which moved with the construction up to the top of the tower. The formwork had four floors and was
supported by 10 ft long, cast aluminum jump beams positioned 8 ft
apart. Each jump beam was secured to the wall by two or four bolts
cast in concrete (Figure 2.2). At the time of the disaster, the thickness
of the wall being cast was 8 in. After each day's casting was finished,
the formwork was raised to the appropriate height by hydraulic jacks
along the jump beams on the outside and on the inside of the concrete
tower. Normally, this system was raised by one-third of its height so
that the weight of the formwork was secured 50% in the previous day's
concrete and 50% in 2-day-old concrete.

In Figure 2.2, three successive pour dates illustrate how the jump
beams were leapfrogged between pours. After the casting on Monday,
April 24, the jump beam was unbolted from the wall and handed up
to workers on the top of the scaffold, who assembled it in place in
anticipation of the next concrete pour. On Wednesday, April 26, the

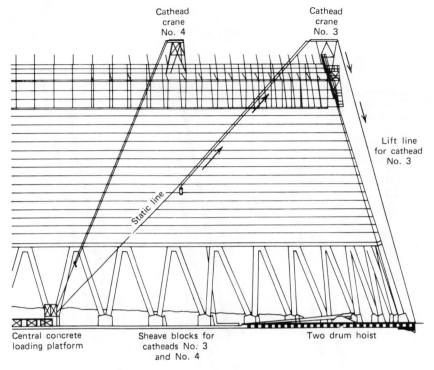

Figure 2.1 Side view of cooling tower.

same process took place on the reverse side of the wall. The configuration of the jump beams and scaffolding on the morning of the collapse is shown in the right part of the figure.

According to OSHA officials (U.S. Department of Labor, 1978), the accident started at cathead crane #4. A laborer working at the base of hoist #4 has just sent up a 2500-lb bucket of concrete when the formwork gave way and the loaded bucket crashed down. The bucket probably was dropped because the supporting frame for the cathead was not secured by bolts. The dynamic load imposed on the structure by the falling concrete bucket was most likely the factor that triggered the collapse. The formwork peeled off the inside of the tower, moving in both directions. The collapse ended after 30 sec, 180° from the starting point. One engineer said, "If everything else was okay, it would have been SOP. You could have dropped the bucket 40 times without anything happening. But when you shock-load the formwork when it is unstable, you start things happening."

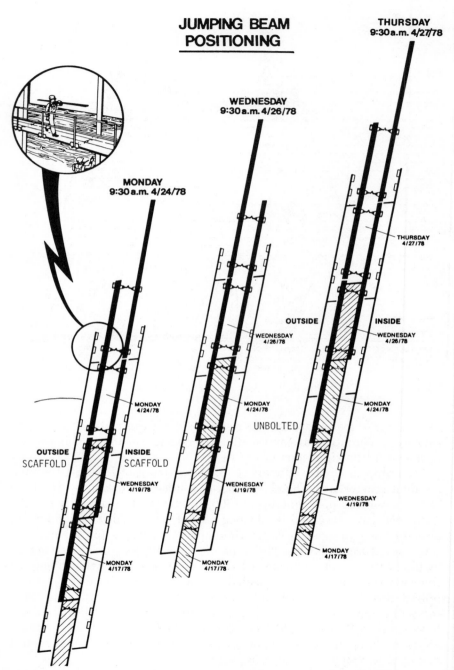

JUMPING BEAM POSITIONING

THURSDAY
9:30 a.m. 4/27/78

WEDNESDAY
9:30 a.m. 4/26/78

MONDAY
9:30 a.m. 4/24/78

THURSDAY
4/27/78

OUTSIDE   INSIDE

WEDNESDAY
4/26/78

WEDNESDAY
4/26/78

MONDAY
4/24/78

MONDAY
4/24/78

MONDAY
4/24/78

UNBOLTED

OUTSIDE   INSIDE
SCAFFOLD   SCAFFOLD

WEDNESDAY
4/19/78

WEDNESDAY
4/19/78

WEDNESDAY
4/19/78

MONDAY
4/17/78

MONDAY
4/17/78

MONDAY
4/17/78

Figure 2.2   Jump beams support formwork and four floors of scaffold structure.

16

2.2 ACCIDENTS HAVE MULTIPLE CAUSES

What can we learn about accidents from the real disaster described above? Although the event that triggered the accident was dropping the loaded bucket of concrete, we cannot say that this, and this alone, caused the accident. There were actually a number of contributing causes, both technical (1–3) and human (4–6) in nature.

1. It is possible that the concrete was substandard and therefore that the bolts that were supposed to support the formwork did not have enough support themselves. The company had not made tests on field-cured concrete to ensure that the concrete had sufficient strength.

2. Some of the bottom bolts (in 2-day-old concrete) on the jump beam had been unbolted before the formwork was subjected to the load of the new pour of concrete. Therefore, the formwork was supported only by concrete from the previous day's casting.

3. The legs of the cathead on top of the formwork had not been anchored and maintained in a way that would support the maximum intended load.

4. The operator of the hoisting equipment was positioned at the base of the tower where he could not see the catheads and had to depend on telephone communication with the workers on top of the formwork.

5. The construction company had patented the formwork design and was very secretive about methods for its use. This placed restrictions on communication between workers and supervisors; there was no real team operation. It is possible that workers did not have a clear idea of how to use the formwork safely.

6. The employees had not been instructed about the danger of premature removal of the components that supported the form scaffold and cathead.

The construction company was cited for all these violations. Because of legal actions brought against the company, the full details of OSHA's accident investigation have not yet been disclosed. Additional citations were issued for the company's failure to inspect all wire ropes at least once a month and to keep proper records of these inspections; for not using a proper equipment-grounding conductor system for portable electric tools; for not equipping the scaffolds with toeboards; and for not providing a wire mesh screen between toeboard and guardrail on scaffolds beneath which people were required to work.

Although the severity of the cooling tower accident is not typical, it illustrates two important general points regarding construction work:

1. The most common type of construction accident is one in which people or objects fall from scaffolds resulting in injuries. Close to 50% of all construction accidents are of this type.

2. There is usually more than a single, isolated cause for any accident. In the major structural accident described above, a number of technical deficiencies and human errors combined to cause the disaster. It is impossible to pinpoint a primary cause.

Singling out one factor as the causal one might provide an oversimplified and misleading explanation. Such a single-factor explanation would divert attention from other, equally important factors. Figure 2.3 diagrams the complexity that is often revealed by accident analysis. Multiple factors acting together are a realistic picture of the complex process of accident causation.

Have we learned anything from the cooling tower catastrophe? Perhaps such an accident gives us a greater appreciation of the importance of providing a safe working environment. It is not easy to specify exactly how a safe working environment might be planned at a construction site. There are many potential causes for accidents in construction work; some involve the machines that are used, some involve worker attitudes and incentives, some involve company policies, work procedures, and so on. All these factors act singly and together in the work environment. Recognizing the complexity of the situation enables us to avoid overly simplistic explanations. Merely regulating the technical aspects of the work, for example, is not likely to solve the total problem of safety in construction.

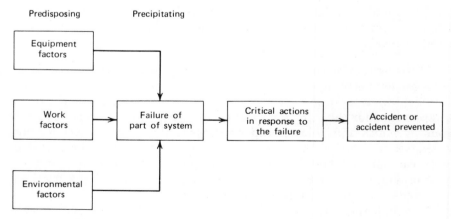

Figure 2.3 Conceptual model of accident-generation process.

2.3 ANALYZING CONSTRUCTION ACCIDENT STATISTICS

The purpose of accident statistics is to identify probable causes of injuries and illnesses so that preventive measures can be taken. Ideally, this goal would dictate the way in which the statistics are collected and analyzed. Unfortunately, accident statistics are often compiled in a way determined by the availability of data rather than by the actual requirements of research (Helander, 1978). Many federal agencies, state agencies, and private companies spend considerable time and effort in producing accident statistics. Several different methods are used for measuring accident occurrences; number of lost-time accidents, number of minor accidents, incidence rate, accident frequency rate, and accident severity rate are all used to gauge how many accidents occur. Typically, these measures are compared with other variables such as age, occupation, nature of accident, type of injury, and "cause" of accident.

Reports on accident statistics rarely contain any conclusions or recommendations. Why, then, are the figures compiled and published (at such cost)? According to Kletz (1976), the statistics are "continued out of habit when their uselessness has become apparent. But the original reason for producing them is the hope that they will show up correlations that will help to prevent future accidents."

In many cases, however, a lack of reliability in the data may prohibit conclusions. Also, different agencies present very different statistics, making overall conclusions difficult to make. For example, the U.S. Department of Labor, Bureau of Labor Statistics, estimated that 1000 construction workers died in 1973 as a result of job-related accidents and illnesses. For the same period, the National Safety Council's estimate was 2900. This discrepancy was caused by the different approaches taken by the two agencies in compiling the accident data. The National Safety Council commented that their approach might not be "as scientific or fancy as the Bureau of Labor Statistics but it is reasonable and the results fit with established data" (Engineering News Record, 1975).

Even after reliable accident statistics have been compiled, they may still be difficult to interpret. There is often no way to correct for differences in exposure between the various jobs in order to make valid comparisons of the dangers involved. For example, working on scaffolds and working in trench excavation produce many major construction accidents. But since no information exists on the relative amounts of time workers spend on scaffolds and in trenches, it is impossible to say how dangerous these jobs are. It may be that, per hour spent working, there are more accidents involved in some other construction tasks. This is like interpreting traffic accident statistics, where it can be seen that people over 65 years of age have very few traffic accidents per 1000 persons. From this statistic one might conclude that older drivers are safer. However, when accidents per million miles driven are calculated, it becomes clear that older drivers have more accidents per mile driven than the average driver.

With these remarks we caution the reader not to accept any accident statistics uncritically, including the ones that follow. There are many uncertainties involved in collecting these data and, since exposure rates are not usually available, conclusions are virtually impossible to draw.

In the United States there are two major publishers of work accident statistics, the National Safety Council and the U.S. Department of Labor. Each organization's method of reporting accidents is different, so their figures cannot easily be compared. Similar trends in the published statistics may exist, however, even when the numbers themselves do not agree.

The U.S. Department of Labor uses "incidence rate" as their only statistic. This expresses the number of health-related incidents during 1 year for 100 full-time workers, and is calculated as follows:

$$\text{Incidence rate} = \frac{N}{EH} \times 200{,}000$$

where N = number of occupational injuries, illnesses, or lost workdays, EH = actual hours worked by 100 employees during the year, and 200,000 is the maximum number of hours worked by 100 employees during the year.

The National Safety Council reports three types of accident statistics: death rate (per million workers), frequency rate, and severity rate of accidents. Frequency rate is the number of disabling work injuries per million hours worked, and severity rate is the number of days charged for accidents per million hours worked.

As mentioned before, construction work is dangerous compared with other occupations. One out of every five construction workers will eventually be seriously or fatally injured at work (Musacchio, 1973). Table 2.1 indicates death rates and disabling injuries for the eight major industries in the United States. It is evident that the construction industry stands out as having the largest number of fatal accidents. Compared with the manufacturing industries, construction has approximately seven times as many fatalities per million workers, and

TABLE 2.1 Death Rates and Rates of Disabling Injuries for Eight Industry Groups in 1977[a]

Industry group	Workers (× 1000)	Deaths per 1000 workers			Disabling injuries per 1000 workers
		1977	1967	% Change	
All industries	90,900	14	19	−18	25
Trade	21,200	6	8	−25	19
Manufacturing	19,600	9	10	−10	27
Service	21,800	8	12	−33	18
Government	15,000	11	13	−12	21
Transportation and public utilities	4,900	33	39	−15	39
Agriculture	3,400	53	68	−22	53
Construction	4,200	60	71	−15	57
Mining, quarrying	800	63	100	−37	50

[a] From the National Safety Council, 1978.

twice as many disabling injuries. In other words, it is much more likely that an accident in the construction industry will result in death or serious injury.

There are many reasons for these differences. In the manufacturing industry the environment and work methods remain essentially unchanged from day to day. At a construction site the environment, the work to be done, and the composition of crews change continually. This makes the overall work situation more complex and difficult to learn; the worker is exposed more often to unforeseen and unaccustomed hazards. The higher rate of fatalities in construction is also due to the more hazardous work methods employed. For example, no other occupation requires the use of scaffolds to such an extent as does construction work. In the United States, 79% of all scaffold accidents occur in construction (U.S. Dept. of Labor, 1979) and approximately 15% of these result in fatalities (Chaffin et al., 1978).

Table 2.1 also compares the death rates between 1967 and 1977: for all occupations, the risk of being killed at work declined over that period. This is also the case over a longer period of time: since the turn of the century, the fatality rates in all industries have declined by 71% (National Safety Council, 1978). This decline reflects several factors: increasing management concern for workers' health, establishment of federal and state safety regulations, and improved work methods. All of these might be said to evidence an increasing concern and maturity in our society.

The incidence rate within the construction industry varies depending on the type of work, the size of the company, and the time of year. In Table 2.2, incidence rates are shown for the different construction specialties (U.S. Department of Labor, 1977b). The most dangerous trades are roofing and sheet metal work. A recent study on roofing accidents (Crisera et al., 1977) showed that the accident rate for roofing

TABLE 2.2   Occupational Injuries in the Construction Industry, 1974 and 1975.[a]

		Incidence Rate Per 100 Full-Time Workers			
	Average employment ($\times 1000$)	Total cases including fatalities		Lost workdays	
Industry group		1974	1975	1974	1975
All construction	3457.0	18.3	16.0	99.8	100.8
General building contractors	1047.9	19.1	16.1	93.2	92.2
Heavy construction contractors	692.5	18.1	16.6	112.7	116.2
Highway and street construction	296.5	15.8	14.8	93.2	102.3
Nuclear energy construction	396.0	19.9	18.1	127.2	127.3
Special trade contractors	1716.8	17.8	15.7	97.9	99.3
Plumbing, heating, air conditioning	413.8	19.2	16.2	90.5	71.4
Painting, paperhanging, and decorating	123.0	11.6	8.6	79.5	78.5
Electrical work	315.0	15.8	14.4	63.6	91.0
Masonry, stonework, plastering	189.1	16.7	14.4	93.3	99.4
Carpentering and flooring	—	14.1	13.4	109.8	94.5
Roofing and sheet metal work	119.9	26.2	23.1	218.4	197.0
Concrete work	—	16.7	15.4	103.5	108.7
Water well drilling	—	18.0	15.2	131.3	118.3
Miscellaneous special trade contractors	—	19.4	17.6	105.5	116.9

[a] U.S. Department of Labor, Bureau of Labor Statistics, 1977b.

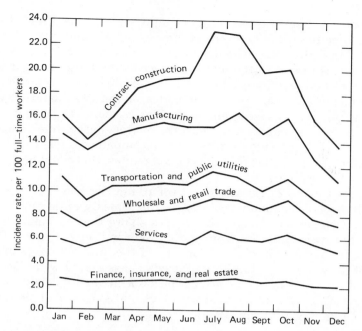

Figure 2.4 Seasonal variation in incidence rates for different trades (U. S. Dept. of Labor, Bureau of Labor Statistics, 1976).

is twice as large as that for underground coal mining (long considered one of the most hazardous occupations). Roofing and some other high-risk construction jobs are discussed more fully in a later section.

Figure 2.4 gives incidence rates for six industries as a function of time of year. Although incidence rates for other industries are fairly stable throughout the year, the construction industry rate shows a large seasonal variation. More construction work is done in warmer weather; there is a seasonal increase in the number of temporary, inexperienced employees. Myers and Swerdloff (1967) observed, "From its low point in February to its peak in August, [the] construction industry . . . adds enough workers to staff the entire motor vehicle manufacturing industry." These workers are likely to be inexperienced and, as such, liable to many more accidents than experienced workers (Vernon, 1954).

A finding common for all types of industries is that the incidence rate is different for companies of different sizes (U.S. Department of Labor, Bureau of Labor Statistics, 1977a). In general, small companies and large ones have lower incidence rates than medium-size companies. There are several factors that might explain this. First, the very large companies may employ safety inspectors. Grimaldi and Simmons (1975) found that the presence of one person whose full-time job is safety seems to reduce the number of accidents. Large companies are often more lavishly financed and can better afford safe, reliable equipment and maintenance. Small companies, on the other hand, may have fewer accidents because their work force is more stable and experienced. Small companies may not take on very large, hazardous jobs. Often, too, the reporting from very small companies is less reliable and less complete than that from larger companies; this could be reflected as a lower accident rate.

Figures 2.5 and 2.6 present causes of all injuries and fatalities in the British construction industry in 1976 (Her Majesty's Stationery Office, 1978a). Comparison among the figures shows that accidents leading to fatalities are different from those producing only injuries. For instance, falls of person account for 50% of the fatal accidents but only 30% of all accidents. Handling of goods is associated with 25% of all types of accidents but is not seen to be associated with fatalities. Comparisons of Figures 2.5 and 2.6 with U.S. statistics reveal similar patterns (State of California, 1977a).

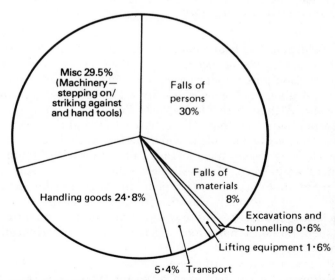

Figure 2.5 Reported accidents in construction in 1976 (total = 34,611) in Britain. (SOURCE: Her Majesty's Stationery Office, 1978a. Reproduced with the permission of the Controller of Her Britannic Majesty's Stationery Office.)

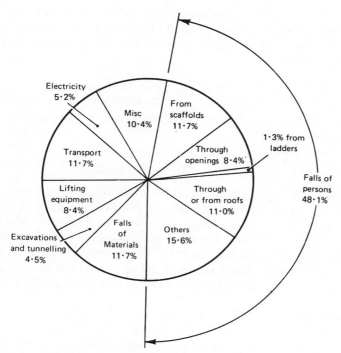

Figure 2.6 Fatal accidents in construction in 1976 (total = 154) in Britain. (SOURCE: Her Majesty's Stationery Office, 1978a. Reproduced with the permission of the Controller of Her Britannic Majesty's Stationery Office.)

2.4 SERIOUS CONSTRUCTION ACCIDENTS AND OSHA

The most frequent causes of accidents in construction are the following:

1. Falls of persons and material (particularly from scaffolds, ladders, and roofs).
2. Collapse of excavations.
3. Misuse or failure of lifting machinery (cranes, hoists, etc.).

These three types of accidents are examined in this section and excerpts from the relevant OSHA regulations are presented.

2.4.1 FALLS OF PERSONS AND MATERIAL

In this subsection we describe some of the frequent causes of these accidents and suggest modifications of the working environment that could reduce the number of accidents of this type. Data from the United States show the same trends as found in Great Britain; Her Majesty's Stationery Office (1978a) reports the following to be the most frequent types of accidents:

1. Falls from working surfaces or gangways.
2. Falls from scaffolds during erection, alteration, or dismantling.
3. Falls from scaffolds due to collapse of part of the scaffold or the whole scaffold.
4. Falls from ladders or stepladders.
5. Falls through fragile roofs or from sloping roofs.

In Table 2.3, an analysis of 98 accidents that occurred between 1972 and 1976 in California is presented. The fatal falls ranged from a few feet to as much as 14 stories. Most falls from roofs, ladders, beams, and trusses involved elevations from 6 to 30 ft, whereas falls from scaffolds had a wider range—from 6 to over 60 ft. These statistics differ from those in Figure 2.6 in one major respect: fatalities due to falls from ladders seem to be more frequent in the United States than in Great Britain.

2.4.2 ACCIDENTS ON SCAFFOLDS

An investigation of 600 scaffold accidents in California (State of California, Department of Industrial Relations, 1974a) showed that approximately 40% of all falls were associated with structural failure, most of which could be attributed to the use of substandard scaffolds. Frequent types of failure included plank broke; plank slipped; scaffold

TABLE 2.3   Analysis of 98 Fatailities Due to Falls from Elevations in Construction in California[a]

Working surface	% Fatalities
Scaffold, staging	22
Roof	21
Beam, truss, joist	15
Ladder	14
Ramp, elevated platform, portable step	9
Building structure	5
Pole	4
Bridge	3
Other elevated surfaces	5

[a] California Department of Labor (1977a).

tilted; scaffold jack, leg, or wheel slipped or came off; scaffold not securely fastened to wall; anchor line broke; or scaffold guardrail broke. For the other 60%, the accident either occurred when a worker lost his balance or the accident reports contained insufficient information to determine the cause of the accident. Statistics gathered in Great Britain, however, indicate that a large number of scaffold accidents occur during erection, alteration, and dismantling of the scaffolds. Unfortunately, these activities are not regulated in the safety and health standards. The accidents that occur with scaffolds today are essentially the same as those that occurred at the beginning of the century. Today, however, new techniques and equipment are available that could reduce the number and seriousness of accidents.

The construction of scaffolds is regulated in detail in the OSHA Safety and Health Regulations for Construction (Federal Register, 1979) as well as in state regulations for construction (e.g., State of California, 1976). The OSHA regulations give detailed instructions for the design of approximately 20 different types of scaffolds. This section reviews some of the general requirements. For details we refer the reader to the above-mentioned sources.

General Requirements for Scaffolds

1. Erection, movement, and alterations shall be under the supervision of a competent person.

2. Scaffolds and their components shall be capable of supporting four times their intended load.

3. All load-carrying timber members of scaffold framing shall be a minimum of 1500-fiber construction-grade lumber.

4. A screen shall be provided between the toeboard and guardrail where persons are required to pass or work under scaffolds.

5. Overhead protection shall be provided for men on a scaffold exposed to overhead hazards.

6. Slippery conditions on scaffolds shall be eliminated as soon as possible.

7. An access ladder or equivalent safe access must be provided.

8. Scaffold planks shall extent over their ends of support not less than 6 in. or more than 12 in.

9. Wire or synthetic or fiber rope used for scaffold suspension shall be capable of supporting six times the rated load.

10. No welding, burning, riveting, or open flame work shall be performed on any staging suspended by means of fiber or synthetic rope.

11. Only treated or protected fiber or synthetic rope shall be used for or near any work involving the use of corrosive substances or chemicals.

As an example of the detailed OSHA regulations, we provide the following summary on construction of wooden scaffolds. For dimensions and complementary regulations, see OSHA Safety and Health Regulations for Construction (Federal Register, 1979).

Construction of Wooden Scaffolds

1. All poles shall be set plumb and bear on a foundation of sufficient size and strength to spread the load from the pole to prevent settlement (see Figure 2.7).

2. Putlogs or bearers shall be set with their greater dimension vertical, long enough to project over the ledgers of the inner and outer rows of the poles at least 3 in. for proper support.

3. Ledgers shall be reinforced by bearing blocks securely nailed to the side of the pole to form a support for the ledger.

4. Diagonal bracing shall be provided.

5. Where wood poles are spliced, the ends shall be squared and the upper section shall rest squarely on the lower section. Wood splice plates shall be provided on at least two adjacent sides and shall be not less than 4 ft in length, overlapping the abutted ends equally, and have the same width and not less than the cross-sectional area of the pole.

6. Platform planks shall be laid tight and where planking is lapped, each plank shall lap its ends support at least 12 in. Where the ends of the planks abut each other, the butt joint shall be at the centerline of a pole. The abutted ends shall rest on separate bearers.

7. Scaffolds shall be guyed or tied to the building or structure at intervals not greater than 25 ft vertically and horizontally.

8. Guardrails and toeboards shall be installed on all open sides and ends of platforms more than 10 ft above the ground.

　　Top rail, 2 × 4 in., 42 in. high
　　Mid rail, 1 × 6 in.
　　Toeboard, 4 in. high.

9. Scaffolds over 60 ft high shall be designed by a qualified engineer.

2.4.3   ACCIDENTS INVOLVING LADDERS

There are more falls from ladders and stepladders than from scaffolds, although the number of fatalities is not as high as for scaffolds. Surveys

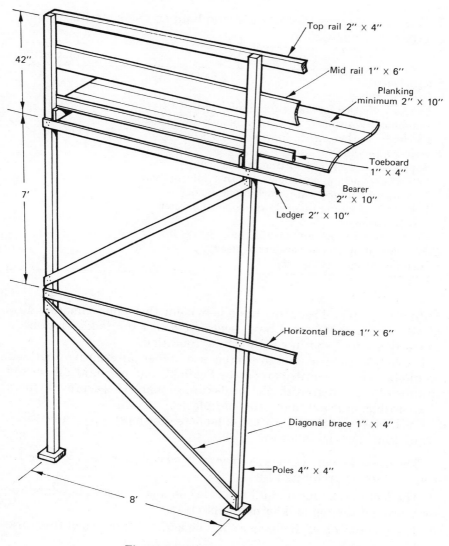

Figure 2.7 Elements of a scaffold.

have shown that the principal cause of ladder accidents is failure to tie or to satisfactorily foot the ladder. Either tie points are not conveniently placed or cost may be an inhibiting consideration in not tying or footing ladders (Her Majesty's Stationery Office, 1978a). An analysis of injuries involving falls from ladders is given in Table 2.4 (State of

TABLE 2.4   Causes for 519 Disabling Injuries Involving Falls from Ladders in Construction

Cause of accident	Injuries (%)
Faulty material	
Ladder slipped	23
Ladder collapsed, broken rung, faulty nut	10
Improperly positioned, unsecured ladder	4
Ladder or worker struck by wind or object	4
Probable human error	
Slipped on ladder, lost footing, missed rung	20
Fell while working on ladder, no other information	18
Became off-balance while working on ladder	11
Fell while climbing up or down ladder, no other information	5
Got electrical shock and lost balance	2
Fall caused by tool or equipment used	2
Insufficient data to classify	2

California, 1977). From this table is evident that approximately 40% of all the accidents could have been avoided if the ladder had been properly secured and in good working condition.

Ladders are cheap and are often used when alternative methods would be safer or more appropriate. Scaffolds and mechanically or hydraulically operated work platforms can, in many cases, provide more comfortable, efficient, and safe workplaces.

OSHA specifies the proper use of ladders. The most important OSHA regulations are the following:

1. The use of ladders with broken or missing rungs or steps or defective construction is prohibited.

2. The feet of the ladder shall be placed on a substantial base and the area around the top and bottom kept clear.

3. The inclination of the ladder shall be about one-fourth of the working length (76°).

4. Side rails shall extend 36 in. above the landing or grab rails be provided.

5. Ladders shall be tied, blocked, or otherwise secured to prevent displacement.

6. Ladders shall not be used in a horizontal position as platforms, runways, or scaffolds.

7. Metal ladders shall not be used for electrical work or where they might contact electrical conductors.

8. The width of a single cleat ladder shall be at least 15 in. but not more than 20 in. between rails at the top.

2.4.4 ACCIDENTS IN ROOFING WORK

Roofing work is the most dangerous activity in the industry. In 1974 there were 11.3 lost workday cases per 100 full-time workers. This exceeds even the rate for underground coal mining (5.7 lost workdays per 100 full-time workers). Several factors account for this:

1. The roofer works in a very irregular environment where only a few days are spent at each job. As a result, safety is typically neglected; for example, ladders and scaffolds are not permanently fastened.

2. Although the materials and application methods are similar from job to job, each building is designed differently and presents a unique set of hazards. Vent holes, skylights, and pipes, for example, are in different locations on each roof. Areas of fragile material may or may not be present.

3. A roof under construction does not provide a firm, strong surface and workers are in danger of stepping in holes or through thin sheeting.

4. The work demands a great deal of lifting; at each new job site employees must unload heavy roofing materials and transport them up to the roof.

5. In roofing work, employees are frequently injured by hot tar and asphalt. Roofing tar must be kept at a temperature of about 450°F (232°C). Under 400°F (204°C), the tar is sticky and difficult to spread; over 500°F (260°C), there is danger of spontaneous combustion (State of California, 1972).

Accidents involving roofing are often serious and account for a large number of fatalities every year. Two types of accidents dominate the fatalities: falls from sloping roofs and falls through roofs made of fragile materials. The categories are about equally frequent.

Accidents resulting from working on sloping roofs have long been recognized as a serious problem. There are methods for reducing the frequency and seriousness of these accidents: guardrails may be installed at roof edges or safety lines or nets may be used. These pose some inconvenience and additional cost, but it must be recognized that worker safety is more important than either.

Falls through fragile roofing material have long been known to be a source of serious injury or death. The fundamental problem is the use of material that, though adequate for its design purpose, cannot support a mass weight. In a majority of cases, these accidents occur on corrugated asbestos or cement sheets. Perhaps because they look so firm and solid, workers become careless in using or working around these materials. The roofing sheets are therefore not only dangerous as unprotected roof edge, but give a wholly misleading sense of security to those who work on them (Her Majesty's Stationery Office, 1978b).

Working with fragile roofing sheets necessitates constant adjustment of crawl boards or ladders. This might, in many cases, be inconvenient and time consuming, causing workers to take chances. The most radical approach to eliminating this type of accident is to require that roofing material be strong enough to support workers and equipment. Since asbestos is also known to cause cancer and other serious diseases, it seems best to abolish its use.

Further research is necessary to develop methods for people to lay roof sheets at exposed edges safely. Mobile scaffolds provide a practical method in many cases, and should be used more frequently. Where mobile scaffolds cannot be used, other methods such as safety nets should be considered.

2.4.5 ACCIDENTS IN TRENCH EXCAVATION

As mentioned previously, trench work is also one of the more dangerous types of construction work. During a 10-year period, 80 workers lost their lives in cave-ins of trenches in California (State of California, 1974b). Suffocation is the chief hazard; nearly 60% of the victims were suffocated. Table 2.5 gives a breakdown of the circumstances of these accidents.

TABLE 2.5 Analysis of 80 Fatalities Resulting from Cave-in of Ditch, Trench, or Excavation in California, 1964–1973

Condition	Fatalities (%)
No shoring or sloping	39
Shoring installed or removed in unsafe manner	19
Inadequate shoring or sloping	14
Working outside shored area	11
Shoring just removed	2
Adjacent pile or wall collapsed or slid	2
Other or not stated	13

The depth of the trenches in which fatalities occurred ranged from only 3 ft (1 m) to 25 ft (8 m). Almost all of these fatalities could have been prevented if construction safety regulations had been followed. These state that all trenches 5 ft or more deep must be adequately guarded against the danger of cave-ins. Yet nearly 40% of the cave-in deaths occurred at job sites where there was no shoring or sloping. The vast majority of these accidents could be prevented by the following precautions (Her Majesty's Stationery Office, 1978a):

1. Shoring and a quantity of suitable timber for wallings, etc., provided on-site and used.

2. Workers trained in the safe installation of such basic equipment. Surprisingly few workers have proper training in timbering and support methods.

3. Adequate and competent supervisors present and attending to changes in conditions.

To make work safer it is also important to implement incentive schemes that discourage workers from taking chances. All too often workers are paid piece rate according to the length of pipe installed. This form of pay does not give workers a paid allowance for erecting necessary shoring.

Another safety problem in trench excavation is that modern excavation machinery digs the trenches to the full depth before support work is installed. When this is the case, it is essential that a safe system of installing support work be adopted so that workers are not endangered during installation. Various methods have been used with success. For example, boxes can be set in the excavation so that men install shoring from inside a protective box (see Figure 2.8). The shoring is installed and tightened against trench sides by means of a spoked wheel. A number of hydraulically and mechanically operated devices are currently on the market that may be installed from the top of a trench to provide side support for shoring.

Work with trench excavation is regulated in detail by OSHA Standards for Construction. Some of the more important regulations are listed below.

1. Banks more than 5 ft high shall be shored, laid back to a stable slope, or some other means of protection provided where employees may be exposed to moving ground or cave-ins. Regulations for angle of repose for open trenches are given in Figure 2.9.

2. Materials used for sheeting and sheet piling, bracing, shoring, and underpinning shall be in good, serviceable condition, and timbers used shall be sound and free from large or loose knots.

3. Determination shall be made of the existence and location of underground installations (sewer, telephone, water, electrical lines, fuel, etc.). Utility companies shall be contacted and advised of proposed work.

4. Trees, boulders, and other surface encumbrances creating a hazard shall be removed or made safe.

5. A competent person shall be provided to make daily inspections of the excavation.

6. Walkways shall be clear of excavated material and shall not be undermined unless shored to carry 120 lb/ft^2.

7. Support systems shall be planned and designed by a qualified person when excavation is in excess of 20 ft in depth, adjacent to structures or improvements, or subject to vibration or ground water. Additional precaution (shoring or braking) shall be taken when trenches are exposed to vibration (railroad, highway, or machinery) or are adjacent to backfill.

Figure 2.8 Protection boxes can be used to provide temporary support until the timbering has been put in the excavation. (SOURCE: Her Majesty's Stationery Office, 1978a. Reproduced with the permission of the Controller of Her Britannic Majesty's Stationery Office.)

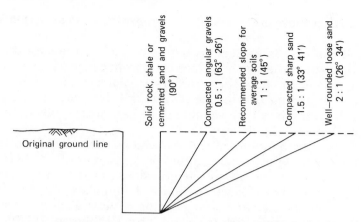

Figure 2.9 Approximate angle of repose for sloping sides of excavations (Federal Register, 1979). Notes: clays, silts, loams, or nonhomogeneous soils require shoring and bracing. The presence of ground water requires special treatment.

2.4.6 ACCIDENTS INVOLVING CRANES

Cranes are the workhorses of the construction industry. One crane operator can now do a job that previously would have required many workers—if it could be done at all. The immediate cause of accidents involving cranes, derricks, and similar equipment generally can be assigned to one of two broad categories (Her Majesty's Stationery Office, 1978a):

1. Lack of adequate training results in many accidents. The most serious of these involve overturning owing to lack of stability or collapse of the crane structure.
2. Structural or mechanical failure of components, resulting from shortcomings in design, cause a substantial number of crane accidents.

According to a Finnish study (Häkkinen, 1978), only 20% of the workers injured in crane accidents are the operators. Many of the operator accidents occurred upon entering or leaving the cab. The majority of crane accidents occurred to workers fastening or loosening loads, or steering loads with their hands during lifting. The most common types of injuries were hands crushed between load and lifting gear, blows from lifting gear or load, and blows from falling loads or lifting tackle. These kinds of accidents can be prevented by proper training of the operator. For instance, it is very important for safe operation that the operator choose the proper lifting gear (Armbrüster, 1976).

A steel chain sling can endure a higher surface pressure than can a wire rope sling; the risk of failure is less when steel chains are used to fasten objects that have sharp corners. In contrast, the sharp bending of a steel wire sling can reduce its strength by more than 50%. To retain full strength, a wire should not be bent more sharply than six times its diameter (Hoffmann, 1974).

Sometimes the handling of the load can be controlled from the operator's cab by use of vacuum lifters or magnets so that there is no need for workers to operate in the hazardous zone near the load. This arrangement has been shown to reduce the number of accidents considerably. The use of automatic gripping devices also increases productivity since it saves both time and labor (Häkkinen, 1978).

A study in West Germany (Miller and Göttling, 1977) of 637 tower crane accidents showed that improper design of the crane can be blamed for many accidents. The most dangerous conditions were found to be erecting, dismantling, or overloading the cranes; these procedures accounted for 43% of all accidents. A majority of these accidents could be avoided given proper design of the machinery. For example, with many kinds of cranes it is necessary to crawl under the machine in order to loosen the bolts that join the various parts. This poses a high risk of worker injury since parts of the structure, once loosened, can fall down. This hazard can be eliminated by use of special bolts which are loosened from the outside of the structure. Some cranes are also difficult to maintain in safe operating condition because joints and bolts are hidden in places where they cannot be easily reached for inspection or lubrication.

Manufacturer's statements such as "99% of all crane accidents are the result of improper use of cranes by callous or unskilled operators" (National Institute for Occupational Safety and Health, 1978) represent a gross oversimplification. A large share of the accidents is actually due to poor design of the equipment that results in difficulties in maintenance, poor visibility, and the like. Some of the unsafe features of older cranes have been eliminated in newer ones: a common feature of machines was that they tended to be unstable. With the development of greater lifting capacity and longer reach, the structure of the crane changed. The problem of a modern crane is less one of stability than of structural strength. This means that now if a crane is overloaded, the crane structure may fail before the crane starts to tip over.

OSHA regulations for the use of cranes are not very detailed, but refer to manufacturer's manuals. Often, however, such manuals provide little (if any) useful information.

2.5 OSHA'S SAFETY REGULATION PROCEDURE

The Occupational Safety and Health Act was adopted by the federal government in 1970 to "assure safe and healthful working conditions to working men and women by authorizing enforcement of the standards developed by the Act; by assisting and encouraging the states in their efforts to assure safe and healthful working conditions; by providing for research, information, education, and training in the field of occupational safety and health; and for other purposes" (U.S. Congress, 1970). This legislation led to the formation of the Occupational Safety and Health Authority (OSHA), the agency charged with enforcing work safety and health regulations.

In 1974, construction safety and health regulations were published. These standards for the most part adopted earlier voluntary regulations issued by the National Safety Council, the American Society for Testing Materials, the American National Standards Institute, the American Institute of Industrial Hygienists, and the National Fire Protection Agency. The regulations are regularly updated; the most recent were published in 1979 (Federal Register, 1979).

Some states independently developed their own standards. In California, for example, the first safety regulations came into effect in 1914. Today, the Cal/OSHA Construction Safety Orders (State of California, 1976) has replaced the OSHA standards. To acquire legal status, state standards must comply with the full intentions of the federal legislation. Since both federal and state standards are very comprehensive, this text provides only a summary of the more important regulations. General requirements are described in this section; some of the major technical requirements have been outlined above (in relevant sections on construction accidents).

2.5.1 EMPLOYEE AND EMPLOYER RIGHTS

An employee has the right to call for a safety inspection by OSHA and to accompany the OSHA compliance officer while he is inspecting the establishment. Employees also have the right to be fully informed of all hazards on the job and of all the actions taken against the employer. Citations for employer violations must be posted prominently at the work site. Employees cannot be discharged or otherwise discriminated against on the job because of exercising any right conferred by OSHA legislation.

The employer has the right to be advised of the reason for inspection by OSHA personnel and to participate in walk-through inspections. The employer can also apply to OSHA for temporary or permanent variance from a standard if it can be proved that the facilities or methods of operation provide protection at least as effective as that required by the standard.

It is important that employers, safety specialists, and management be fully acquainted with OSHA regulations. The standards can actually be used in a positive way to facilitate cooperation and discussion between management and labor in their mutual efforts to reduce job hazards.

2.5.2   COMPLIANCE INSPECTIONS

Inspections are made by compliance safety and health officers and industrial hygienists. To ensure that they will satisfy OSHA's qualification requirements, each compliance officer receives special training before doing work in the field. When the compliance officer comes to a workplace, he or she presents credentials and asks to meet the employer. The officer then informs the employer of the reason for the visit and describes in general terms the procedures for the inspection. These include a review of safety and health records, employee interviews, a walk-around inspection of the facilities, and a closing conference. The employer is given copies of applicable laws and safety and health standards, along with a copy of an employee's written complaint (if one is involved). The employee's name is withheld upon request.

The compliance officer asks the employer to designate representatives for the walk-around inspection. If there is a union at the workplace, an employee is asked to serve as a union representative during the walk-around inspection. If there is no employee group, the compliance officer discusses conditions with individual employees. During the walk-around, the compliance officer takes notes of conditions and discusses them with both the employees' and employer's representatives. The officer may also take air samples and photographs but is obliged by law to protect trade secrets and company security.

After the walk-around inspection, the compliance officer holds a closing conference with the employer, discussing the inspection and reviewing possible violations. A report is then prepared for the OSHA area director, who decides (with help from his or her supervisors) whether citations should be issued and what penalties, if any, should be imposed.

If the compliance officer discovers an imminently dangerous situation, the employer and employees are informed of the danger and of the fact that the officer is recommending to the Secretary of Labor that a civil action be initiated in a U.S. District Court. The Secretary then determines whether the danger warrants court action. If the decision is not to seek a court restraining order, the employees still have the right to bring court action compelling the Secretary to do so.

2.5.3 VIOLATIONS

If the compliance officer has reason to believe that the employer has violated the OSHA regulations, a citation may be issued to the employer. There are several types of violation; from least to most serious, these are *de minimis*, nonserious, serious, willful, repeated, and imminent danger. For the most serious types of violation fines up to $10,000 can be issued. For the least serious violations, fines of $1000 are issued. If an employer disagrees with a citation or the penalty, he can request an informal meeting with the area director to discuss the case. He can also decide to contest the citation legally.

In case this all sounds mighty hard on the employer, consider one of OSHA's historical predecessors. The world's first building code was devised by King Hammurabi in 2200 BC for Babylonian building contractors. Figure 2.10 illustrates what they were up against.

2.5.4 RECORDING AND REPORTING REQUIREMENTS

According to OSHA regulations, there are certain records that must be kept by the employer: work injuries and illnesses resulting in fatalities; lost workday cases other than fatalities; and cases without lost workdays that result in transfer to another job, termination of employment, or medical treatment. These must be recorded in a log of occupational injuries and illnesses by every employer with eight or more employees. In addition, the employer is required to compile an annual report of this information.

Special records are also required for a number of types of equipment and hazards:

1. *Scaffolding.* Drawings and specifications of tube and copper scaffolds must be available for the OSHA inspector.
2. *Power Platforms.* Logs of tests and inspections must be kept.

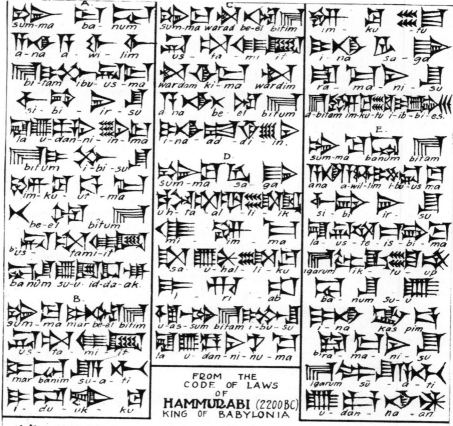

A. If a builder build a house for a man and do not make its construction firm and the house which he has built collapse and cause the death of the owner of the house — that builder shall be put to death.

B. If it cause the death of the son of the owner of the house — they shall put to death a son of that builder.

C. If it cause the death of a slave of the owner of the house — he shall give to the owner of the house a slave of equal value.

D. If it destroy property, he shall restore whatever it destroyed, and because he did not make the house which he built firm and it collapsed, he shall rebuild the house which collapsed at his own expense.

E. If a builder build a house for a man and do not make its construction meet the requirements and a wall fall in, that builder shall strengthen the wall at his own expense.

Figure 2.10   This is a copy of the oldest written building code (translation from Harper, 1904).

3. *Man Lifts.*   Logs of periodic inspections must be kept.

4. *Flammable and Combustible Liquids.*   Inventory records on Class 1 liquid storage tank contents must be kept.

5. *Portable Fire Extinguishers.*

6. *Overhead and Gantry Cranes.*   Reports of monthly rope inspections must be made.

7. *Crawler-Locomotive and Truck Cranes.*   Records of monthly inspections of critical items such as ropes, brakes, and crane hooks are required.

8. *Derricks.*   A monthly rope inspection record is required.

9. *Welding, Cutting, and Bracing.*   Records must be kept of periodic inspection of resistance welding equipment.

2.5.5   POSTING

Employers are required to post OSHA information prominently at the work site. This information includes the OSHA poster entitled "Safety and Health Protection on the Job," the annual summary of occupational injuries and illnesses, information warning of the danger of exposure to toxic or dangerous materials, and citations from OSHA officials.

2.5.6   OSHA'S INSPECTION PRIORITIES

There are approximately 5 million workplaces in the United States. Clearly, not all of these can be inspected frequently, and some will probably never be inspected by OSHA. OSHA has established priorities for inspections by OSHA officials:

1. Sites of catastrophic and other fatal accidents.
2. Investigation of employee complaints.
3. Special-emphasis inspection programs.
   a. Target industries.
   b. Target health hazards.
4. A sample of all types of workplaces in all sections of the United States.

In spite of the fact that OSHA is concerned with all industries, the majority of OSHA inspections are actually made in the construction industry. For example, California currently employs approximately 200 safety inspectors. Up to 45% of all inspections to date have been directed at the construction industry, even though construction employs only 6% of the work force.

2.5.7   DOES THE OCCUPATIONAL SAFETY AND HEALTH PROGRAM WORK?

The OSHA program has received much criticism from the construction industry. Common complaints have been that safety standards are too technical and do not recognize that as many as 70% of accidents may be due to human rather than technical failure (e.g., Surry, 1968). OSHA legislation seems to direct safety investments into an area where the rate of return is comparatively small. In addition, contractor's spokesmen claim that since the OSHA Act changed accident record-keeping techniques, it is difficult to make meaningful comparisons of pre-OSHA and post-OSHA statistics. Representatives of employers' associations believe that a comparative study would prove that OSHA has not been very effective: "Members of our association have been keeping accident records for their own information. Despite several years of federal enforcement, we have found no significant change in accident rates" (Schmuhl, 1978). However, AFL-CIO representatives disagree to some extent: "I grant that we still have a long way to go, *but at least fatalities are down*. I think federal enforcement has made the whole industry more safety conscious and that we are on the brink of significant progress" (Lapping, 1978). Neither viewpoint is supported by convincing statistics.

Even if statistics were available for comparison, it is not clear that it would be possible to isolate the influence of the OSHA Act. The general reduction in accidents for the last few years might reflect increased unemployment (inexperienced workers are laid off first), fluctuations in weather, changes in building methods, or other, unknown factors. Although the OSHA legislation is aimed at reducing technical hazards, one should not overlook the possibility that the greatest impact of the regulations may be to make people more safety conscious and to encourage attitudes that promote a safe working environment. Such a side effect would be a great accomplishment since, where a positive attitude toward safety exists, accidents may be reduced drastically. This is further discussed in Section 2.7.

2.6   ECONOMICS OF CONSTRUCTION ACCIDENTS

In addition to the human costs, construction accidents have substantial economic consequences for the contractor. The direct costs of accidents are fairly easy to assess, such as costs of workmen's compensation insurance premiums. There are always indirect costs that are more difficult to quantify. These are generally estimated to be much higher than the direct costs.

2.6.1 DIRECT COSTS

The purpose of workmen's compensation insurance is to pay for health care and provide economic compensation to workers who are injured at work. Workmen's compensation benefits include the following:

- Medical and hospital care.
- Economic compensation for lost income, up to a specified maximum.
- Compensation for permanent disability.
- Costs for rehabilitation.
- Payments to dependents of deceased workers.
- Burial expenses.

Under the workmen's compensation system, the employer accepts a certain, limited liability for the accident without admitting any fault. At the same time, the worker surrenders his right to sue for unlimited damages and accepts the limited liability of the employer.

The cost of workmen's compensation premiums typically range between 1 and 15% of wages, depending on the hazardousness of the work. The cost of workmen's compensation is a major cost item for contractors, and is comparable in amount with the contractor's calculated profit (Levitt, 1975). The cost of workmen's compensation is also affected by the accident experience of the individual contractor. A construction company with few accidents pays a low premium, and vice versa. This adjustment of insurance rates is called "experience modification." The contractor must live with his safety record of the past 3 years. Even when he has a relatively accident-free year he must wait for some time before this is reflected in lower premium costs.

The total cost to the industry is very great. Assume that there were 5 million construction workers in the United States in 1979 with an average hourly earning of \$9.20 (U.S. Department of Labor, 1979). From Table 2.3 their lost workday incidence rate is 100. The actual number of lost workdays (N) can be calculated:

$$N = \frac{5 \times 10^6 \times 100}{100} = 5{,}000{,}000$$

Assuming a disability payment of half the wage, the workers' lost
wages can be estimated at $183 million. Given an average weekly
salary of $337, the annual wages paid in the construction industry can
be assessed at $84.2 billion. Since the workmen's compensation pre-
mium is approximately $5 per $100 payroll (Levitt, 1975), the total
cost of workmen's compensation is estimated at $4.2 billion.

2.6.2 INDIRECT COSTS

The various cost components of construction accidents are shown in
Table 2.6. It can be seen that the indirect costs are much higher than
the direct costs. Indirect costs include the cost to contractors for un-
insured losses, psychological costs to the worker and/or his family for
disability or death, and costs to federal and state authorities for maintaining OSHA inspection and administration. The indirect costs are
typically estimated by insurance companies to be between four and ten
times the direct costs of accidents (Levitt, 1975). A conservative estimate would be that the indirect costs are twice as large as the direct
costs, or approximately $8.8 billion/year.

A contractor has substantial indirect costs for accidents. A serious
accident has a large impact on workers; serious injuries and fatalities
make co-workers feel dispirited and little motivated to work. A decrease in productivity can be expected for at least 1 week following the
accident (Williams, 1979). Construction accidents often involve much
more serious losses of material, equipment, and/or structures than

TABLE 2.6   Cost of Construction Accidents 1979

Costs	Incurred by	Billion $
Direct costs		
Loss of wages	Workers	0.2
W-C premiums	Contractor	4.2
Indirect costs		8.8
Replacement of worker	Contractor	
Lower productivity	Contractor	
Repair of equipment and structures	Contractor	
Administrative costs	Contractor	
Psychological costs of fatalities and disabilities	Workers	
OSHA legislation and enforcement	Federal and state authorities	
	Total	13.2

accidents in other industries. The replacement/rebuilding can involve great expense. The costs for replacing a journeyman are conservatively assessed at $1000; this includes the cost of hiring, training, and lower initial productivity. Until the injured worker can be replaced, interruption of planned work lowers productivity. Finally, there are increased administration costs involved in litigation and reporting of the accident, etc.

All the costs incurred by contractors are eventually passed on to consumers in the form of higher housing costs, higher utility bills, and higher costs of manufactured products. Accident reduction, particularly in construction, should clearly have a high priority in overall national economic interest. To everybody involved in the construction industry it should be clear that accident costs are close to 10% of the total of all building costs, or approximately 1% of the gross national product in the United States. [Similar estimates have been derived by Levitt (1975).] This should certainly provide incentive enough for accident-reduction programs that would enable contractors to lower their overhead and for federally sponsored research to develop means for making work safer.

2.7 MAKING CONSTRUCTION WORK SAFER

We began this chapter with the caution that it would be impossible to describe everything there is to know about construction accidents. We have only been able to skim the field and examine a few of the most generally relevant features. We have surveyed OSHA regulations and legislation on construction safety; we provided some accident statistics that describe the extent and nature of the problems. Most important, perhaps, is that we have conveyed the idea that accidents are difficult to analyze and, without detailed study, it is virtually impossible to isolate the true causes of accidents.

What, then, can a construction manager do to increase safety? Most construction managers believe that it is important to attempt to do so. A study by Levitt (1975) showed that top managers have specific reasons for promoting safety: some are motivated by humanitarian concern, some by the high cost of workmen's compensation, and some by a desire to maintain a good safety image.

The present understanding of the safety problem makes it possible to identify two complementary methods or approaches for increasing construction safety. The first approach is to promote technical types of regulation (such as OSHA) to ensure a basic level of equipment and procedural safety for workers. The second approach involves designing and implementing programs that directly affect company attitudes toward working safely.

2.7.1 TECHNICAL REGULATIONS

The OSHA regulations frequently quoted in the preceding text provide the framework for promoting safety from a technical point of view. The OSHA Act has focused much of the construction industry's safety effort on compliance with safety and health standards. In fact, safety personnel often complain that the task of keeping up with the frequent changes in legislation severely reduces the time available for working directly on safety problems. However, there are a number of positive features to the legislation. It has been effective in reducing some construction work hazards. Even more important, it has made workers and management more aware of potential hazards and of ways to reduce these.

A safety program should, as a minimum, provide the following:

1. A safety introduction program for new employees.
2. Weekly on-site meetings in which relevant safety issues and problems are discussed. This is required by law in some states.
3. Inspection of the workplace according to OSHA standards. Table 2.7 is a job safety checklist that can be used for this purpose.

TABLE 2.7   Job Safety Checklist[a]

Job #____ Job Name _____ Job Superintendent _____ Date _____
Mark __ for OK; __ for not applicable; and __ for correction needed
1. *Hard hats:* Are they being worn by EVERYONE on job sites?   _____
2. *Gas cylinders:* Secured vertically and valves capped when not in use?   _____
3. *Ladders:* Must have no broken or loose steps   _____
   Must be secured at top   _____
   Must extend at least 36 in. above uppermost landing   _____
4. *Housekeeping:* Are work areas, stairways, and passages clear?   _____
   Are trash containers available and used?   _____
5. *Floor, roof and wall openings, open-sided floors and platforms:*
   Are top rails, intermediate rails, and toeboards in place?   _____
   Are top rails 42 to 45 in. above deck?   _____
   Are toeboards 6 in. high?   _____

6. *Illumination:* Is a 100-W bulb located not more than 10 ft above every stair landing? _____
   Are bulbs recessed or guarded to prevent accidental contact? _____
   Are work areas safely lighted? _____
7. *Crane, hoists, and derricks:* Is a fire extinguisher at operator station? Are cables inspected regularly? _____
8. *Forklifts:* Is the rated capacity posted and clearly visible to the operator? _____
   Is the ground level enough for safe operation? _____
9. *Trenches and excavations:* Are sides sloped? _____
   If not, is shoring in place as required for 5 ft depth or more? _____
10. *Power saws:* Are guards in place and operating properly? _____
    Is there an anti-kickback device installed? _____
    Is it used while ripping? _____
11a. *Scaffolds, fixed:* Are they on a good, solid base? _____
    Are there at least two scaffold planks in every span (at least three planks on 4 ft wide metal scaffolds)? _____
    Do planks overlap ledgers at least 6 in.? _____
    Is there no more than 8 in. space horizontally between plank and guardrail? (May be up to 16 in. if mid rail is used.) _____
    Are guardrails on ends as well as sides of scaffolding? _____
11b. *Scaffolds, rolling:* Are planks cleated so they can't slip off? _____
    Are guardrails, mid rails, and toeboards all around? (Required if scaffold is higher than 6 ft.) _____
    Are wheels pinned to scaffold? Height cannot exceed three times the minimum base dimension. Wheels or casters must be lockable. _____
    Platform must cover entire scaffold width. _____
12. *Personal protective and life saving gear:* Are jackhammer and sandblast men wearing ear plugs? _____
    Are welders wearing dark hoods or dark glasses? _____
    Are respirators, safety belts, and lifelines worn where needed? _____
13. *First aid:* Are supplies available on site? _____
    Are doctor and ambulance phone numbers posted? _____
14. *Fire protection:* Is there at least one extinguisher or water barrel with two buckets on every floor? Horizontal distance to this equipment cannot exceed 100 ft. _____
15. *Temporary lights and other electrical cords:* Must be protected against damage from vehicles, door pinching, etc. _____
16. *Stairways:* Are handrails in place 30 to 34 in. above tread nosing? _____
    Are there a guardrail, mid rail, and toeboard around all stairwells, balconies, landings, and porches? _____
17. Are air hoses all safety-chained or wired? _____

a Courtesy of Max Williams, Safety Director, Williams and Burrows, Inc., General Building Contractors, Belmont, California.

2.7.2 COMPANY ATTITUDES

It has been estimated that approximately 70% of the accidents in the construction industry involve human error in some way; human error may be a primary or contributing cause. Unfortunately, the kind of technical approach taken by OSHA cannot deal with elements of human error in accident causation. One approach that may be useful in counteracting this problem concerns attitudes and motivation. A research group at Stanford University investigated the effect of manager safety motivation on accident frequency. The study of 23 construction companies (Levitt, 1975) showed that top managers can reduce accident rates and costs significantly by doing the following:

1. Knowing the safety records of all field managers and giving safety record some priority in evaluating field personnel for promotion or salary increases.

2. Communicating about safety on job-site visits in the same way they communicate about costs and schedules.

3. Using a cost accounting system to encourage safety by allocating safety costs to company accounts and allocating accident costs to projects.

4. Requiring detailed work planning to ensure that equipment and materials needed to perform work safely are on hand when required.

5. Insisting that newly hired employees receive training in safe work methods.

6. Discriminating in the use of safety awards. Data suggest that safety awards for workers should be based on fewest minor injuries rather than on the number of lost-time accidents. Safety awards for managers should be bonuses based on minimizing lost-time accidents or insurance claim costs.

7. Making effective use of the expertise of safety departments where these exist.

Levitt concluded that companies that observed these rules reduced accidents by 75% compared to companies that did not. The reduced accident rate, of course, lowers the cost of workmen's compensation. For a construction company with 500 employees, such a reduction would mean savings in workmen's compensation premiums of more than $100,000. This would certainly justify management attention and even pay for the employment of a safety engineer.

The effect of middle management and foreman attitudes on safety has also been investigated (Hinze, 1976; Samuelson, 1977; Andriessen,

1978). These reports confirm the importance of company attitudes on safety records. Andriessen points out the following:

1. Management attitudes have more influence on safety than do foreman attitudes or peer group attitudes.

2. Workers will work more safely for a supervisor who is seen as someone who respects his men and their contribution, and who is stimulated by a definite company policy on safety. It can be assumed that a supervisor who is positive about safety will also give more information and advice on safety to show his personnel that working safely is important.

3. Groups that are well coordinated in their work and in which there are few misunderstandings work more safely. In such groups there is a positive atmosphere and members encourage each other to work both efficiently and safely.

4. The personality characteristics of the workers have little influence on their safety records.

5. Working on a piece-rate basis increases the number of accidents considerably compared to other kinds of incentives. A Swedish study by Fredin et al. (1974) confirms this finding.

2.7.3 THE NEED FOR HUMAN FACTORS RESEARCH

Although construction work methods have changed radically during this century, many of the hazards that cause serious accidents have remained the same. This can be demonstrated by comparing current statistics to studies performed in Great Britain at the beginning of the century. One study from 1907 pointed to the dangers of working with scaffolds and, in 1912, the then Home Secretary Winston Churchill, formed a special committee to investigate accidents in construction excavations (Her Majesty's Stationery Office, 1978a). It seems that the construction industry has done little to reduce these hazards in spite of the fact that they have been so long recognized.

Perhaps one reason why the well-known dangers have not been lessened is that research has concentrated on the technical aspects of accident causation rather than focusing on the human causes that may prove more important. The latter approach has been used to great benefit in traffic safety studies. One study performed at Indiana State University (Treat et al. 1977) found that approximately 70% of traffic accidents involve human error, not mechanical or design failure of the car.

Treat et al. (1977) found that the most common causes of human error in traffic accidents were improper lookout, excessive speed, inattention, internal distraction, incorrect driving techniques, false assumption, and improper maneuver. This kind of information can be attained only by in-depth analysis of the situation, the vehicle, and the driver by people versed in the techniques of such analysis and knowledgeable about psychology and the driving task.

Similar in-depth investigations have not been done for construction work or many other industries. This makes it difficult to guess how much human error is involved in accidents in these industries; it makes it impossible to suggest remedies for the problem. For example, it has been shown that alcohol intoxication causes human error in about 50% of all fatal traffic accidents. How large is the alcohol involvement in industrial accidents? Surry (1968) stated that "no North American study has investigated the involvement of alcohol in work accidents. This is presumably due to the belief that drinking is well controlled in industry. However, the damning evidence of alochol involvement in traffic accidents suggests that it would be worthwhile to take blood samples of industrial accident victims and compare them to fellow workers. Until it is proven that very little drinking occurs at or before work, the possibility of the effects on accident liability must be borne in mind." Surry noted that two French studies indicated involvement of alcohol in 10% of all industrial accidents and 29% of those requiring hospitalization.

It is clear that, currently, very little information exists on the contribution of human error to construction accidents. Until the appropriate research is done, it is impossible to devise specific procedures for avoiding these kinds of errors. The current technical approach to safety, exemplified by the OSHA regulations, might be the best way to go, at least until we know more.

REFERENCES

Andriessen, J. H. T. H. Safe Behaviour and Safety Motivation. *Journal of Occupational Accidents*, **1**, 363–376, 1978.

Armbrüster, A. Auswahl eines Anschlagmittels unter sicherheits-technischen Aspekten. *Deutsche Hebe- und Fördertechnik*, **22**(12), 37–38, 1976.

Chaffin, D. B., Miodonsky, R., Strobbe, T., Boydstrum, L. and Armstrong, T. *An Ergonomic Basis for Recommendations Pertaining to Specific Sections of OSHA Standard 29CFR part 1910, subpart D—Walking and Working Surfaces*. Ann Arbor, Mich.: University of Michigan, Department of Industrial Engineering, 1978.

Crisera, R. A., Martin, J. P., and Prather, K. L. *Supervisory Effects on Worker Safety in the Roofing Industry*. San Pedro, Calif.: Management Sciences Co., December 1977.

Engineering News-Record. Construction Fatalities. *Engineering News-Record*, **46**, January 16, 1975.

Federal Register. Safety and Health Regulations for Construction. *Federal Register*, **44**, 8575–8858, February 9, 1979.

Fredin, H., Gerdman, P., and Thorson, J. Industrial Accidents in the Construction Industry. *Scandinavian Journal of Social Medicine*, **2**, 67–77, 1974.

Grimaldi, J. V. and Simmons, R. H. *Safety Management*. Homewood, Ill.: Richard D. Irwin, 1975.

Häkkinen, K. Crane Accidents and Their Prevention. *Journal of Occupational Accidents*, **1**, 353–361, 1978.

Harper, R. F. *Hammurabi, King of Babylonia*. Chicago: The University of Chicago Press, 1904.

Helander, M. *The State of the Art of Ergonomics in the Construction Industry* (Tech. Rep. 2707). Goleta, Calif.: Human Factors Research, Inc., 1978.

Her Majesty's Stationery Office. *The Construction Industry: Health and Safety 1976*. London: 1978a.

Her Majesty's Stationery Office. *One Hundred Fatal Accidents in Construction*. London: 1978b.

Hinze, J. *The Effect of Middle Management on Safety in Construction* (Rep. 209). Palo Alto, Calif.: Stanford University, the Construction Institute, 1976.

Hoffmann, F. Sicheres Anschlagen von Lasten. *Reihe Materialinfluss in Betrieb*. Düsseldorf: VDI-Verlag Gmbh, 1974.

Kletz, T. A. Accident Data—The Need for a New Look at the Sort of Data that are Collected and Analyzed. *Journal of Occupational Accidents*, **1**, 95–105, 1976.

Lapping, J. Personal communication. 1978.

Levitt, R. E. *The Effect of Top Management on Safety in Construction* (Rep. 196). Palo Alto, Calif.: Stanford University, Department of Civil Engineering, the Construction Institute, 1975.

Miller, H. and Göttling, H. Schäden an Turmdrehkranen, deren Ursachen und Möglichkeiten zu ihrer Verhütung. *Der Maschinenschaden*, **45**(2), 53–64, 1977.

Musacchio, C. The Construction Industry's Struggle for Safety. *Occupational Hazards*, 33–35, October 1975.

Myers, R. J. and Swerdloff, S. Seasonality and Construction. *Monthly Labor Review*, 1–8, September 1967.

National Safety Council. *Accident Facts, 1978 Edition*. Chicago: 1978.

National Institute for Occupational Safety and Health. *Health and Safety Guide for Crane Operators*. Cincinnati, Ohio: 1978.

Prather, K., Crisera, R. A., and Fidell, S. *Behavioral Analysis of Workers and Job Hazards in the Roofing Industry*. Los Angeles: Theodore Barry & Assoc., June 1975.

Samuelson, N. M. *The Effect of Foremen on Safety in Construction* (Rep. 219). Palo Alto, Calif.: Stanford University, Department of Civil Engineering, the Construction Institute, 1977.

Schmuhl, A. Personal communication, Washington, D.C.: 1978.

State of California. *Work Injuries in Roofing and Sheet Metal Work*. San Francisco: State of California, Department of Industrial Relations, 1970.

State of California. On-the-Job Scaffold Accidents, California. *Work Injuries in California Quarterly*, **IAQ-36**, 3–7, 1974a.

State of California. Work Fatalities Resulting from Cave-in of Ditch, Trench or Excavation. *Work Injuries in California Quarterly,* **IAQ-35,** 1–7, 1974b.

State of California. *Cal/OSHA Construction Safety Orders.* Los Angeles: Building News, Inc., January 15, 1976 (5th printing).

State of California. Work Fatalities Involving Rollover of Earth-moving Equipment. *Work Injuries in California Quarterly,* **IAQ-42,** 1–7, November 1976.

State of California. Falls from Elevations in Construction in California. *Work Injuries and Illnesses in California-Quarterly,* **IAQ-44,** 1–10, November 1977.

Surry, J. *Industry Accident Research.* Toronto: Labor Safety Council, Ontario Ministry of Labor, 1968.

Treat, J. R., Tumbas, N. S., McDonald, S. T., Shinar, D., Hume, R. D., Mayer, R. E., Stansifer, R. L., and Castellan, N. J. *Tri-level Study of the Causes of Traffic Accidents* (Rep. DOT-HS-034-3-535-77 TAC). Bloomington, Ind.: Indiana University, March 1977.

U.S. Department of Labor, Bureau of Labor Statistics. *Occupational Injuries and Illnesses in the United States by Industry, 1974.* (Bull. 1932). Washington, D.C.: 1976.

U.S. Department of Labor, Bureau of Labor Statistics. *A Guide to Evaluating Your Firm's Injury and Illness Experience, 1975* (Rep. 491). Washington, D.C.: 1977a.

U.S. Department of Labor, Bureau of Labor Statistics. *Chartbook on Occupational Injuries and Illnesses in 1975* (Rep. 501). Washington, D.C.: 1977b.

U.S. Department of Labor, Bureau of Labor Statistics. "Survey of Scaffold Accidents." Unpublished report. Washington, D.C.: 1979.

U.S. Department of Labor, Office of Information. *OSHA Announces Findings of Investigation into West Virginia Tower Collapse* (USDL 78-525). Washington, D.C.: June 8, 1978.

U.S. Congress. *Occupational Safety and Health Act of 1970. Public Law 91-596.* Washington, D.C.: U.S. Government Printing Office, 1970.

Williams, M. Personal communication, 1979.

CHAPTER 3

PHYSICAL HEALTH HAZARDS IN CONSTRUCTION

CARL ZENZ, M.D., Sc.D.

Milwaukee, Wisconsin

3.1 OCCUPATIONAL EXPOSURE TO NOISE: PHYSIOLOGICAL ASPECTS OF HEARING AND HEARING LOSS

It is well known that exposure to intense noise may cause impairment or loss of hearing and interference with communication. Since the early 1950s, this problem has received intensive scientific study, yet many questions remain unsolved. More data on human hearing and noise exposure must be analyzed before firm conclusions can be drawn concerning the relationships of long-term occupational noise exposures and degree of hearing loss in workers. A brief discussion of the physiology of hearing in man follows.

The ear consists of three parts, the external, middle, and inner ear (Figure 3.1). The external ear is bounded by the ear flap and the ear drum (tympanic membrane). The external canal is about 25 mm long and has an S-shaped curve directed inward, upward, and backward. The external ear and canal serves as a conducting channel for sound, acting somewhat like a megaphone.

The middle ear is bounded by the inner layer of the eardrum, and internally by the mucous membrane covering the bony wall of the inner ear. Within the middle ear cavity there are three tiny ear bones—the hammer, anvil, and stirrup—and two small muscles, the stapedius

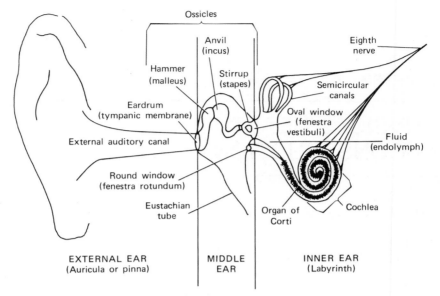

Figure 3.1   Schematic diagram of anatomy of the ear.

(which is attached to the stirrup or stapes) and the tensor tympani (which is inserted on the hammer or malleus). The middle ear presents four openings—the oval and round windows, the opening to the Eustachian tube, and the opening into the mastoid cavity. The oval window is fitted with the flattened bony foot plate of the stirrup. The window is held in place by a fibrous membrane that encircles the bony foot plate and acts as a hinge permitting limited movement of the stirrup. Thus movement originating at the eardrum transmits vibrations from the air (sound waves) to the inner ear. The round window is a tiny opening located just below and behind the oval window. It is covered by a thin membrane resembling that of the eardrum.

The middle ear has three main functions:

1. *Transmission of Sound.*   The hammer, anvil, and stirrup are responsible for the efficient transmission of the sound vibrations in the air column within the auditory canal across the middle ear to the fluid contained in the inner ear.

2. *Protection for the Inner Ear.*   The stapedius and tensor tympani muscles act as a protective device by reflexively tensing the drumhead and ear bones, thereby stiffening the system which carries the sound

vibrations across the middle ear. This reflex acts to reduce the shock to the inner ear in case of sudden, low-pitched sounds.

3. *Maintenance of Pressure.* The Eustachian tube which connects the middle ear with the posterior nasal cavity allows air to enter and leave the middle ear cavity, thereby maintaining equal pressure on both sides of the eardrum. This function is most evident with changes in altitude.

The inner ear consists of two main parts, the semicircular canals (these contain the balance mechanism) and the cochlea. The cochlea is a snail-like tube with two and one-half turns. It contains a system of membranous canals within which lies the end organ of the hearing nerve. One canal travels up the spiral from the oval window and, without a break in continuity, turns on itself and returns to the round window. The second canal is essentially a closed tube located within the first. It is the closed tube that contains the auditory nerve ending, called the organ of Corti.

Since the membranous canals contain fluid, they function as a hydrodynamic system. Motion pictures of the middle ear demonstrate that inward movement of the stirrup produces an outward movement of the round window membrane. This allows wave forms to travel along the fluid of the double canal, which produces movements of the auditory nerve hair cells. These movements trigger nerve impulses which are carried to the brain for interpretation.

In general, the young human ear can distinguish frequencies between 20 and 15,000 Hz. The frequencies important for speech intelligibility range between 500 and 2000 Hz. The speech area is further divided into the vowel and consonant frequencies. The vowel frequencies lie between 500 and 1000 Hz and the principal consonant frequencies lie between 1000 and 2000 Hz. The ear is most sensitive to sound between 1000 and 4000 Hz and becomes less sensitive to sound frequencies above and below that range. Acuity for the higher frequencies, particularly those above 4000 Hz, diminishes with age: Few people over 50 years of age can hear sounds of frequencies greater than 12,000 Hz.

3.1.1 CAUSES OF HEARING LOSS

Partial or total deafness may affect either or both ears. Hearing loss may be caused by infections, obstructions in the ear, or physical agents (including noise). Infections of the external ear may cause swelling of the external auditory canal and result in obstruction and hearing loss. Chronic infections of the middle ear may cause mild to severe conduction hearing loss. Acute infections may heal completely. Infection is generally caused by infection elsewhere in the body, usually in the respiratory tract.

Wax in the external canal is a frequent cause of hearing loss owing to obstruction of the canal. There is a normal secretion of wax in all people but the quantity and consistency show much individual variation.

Foreign bodies in the ear canal or, rarely, in the middle ear, cause infection with damage to the entire ear. Insects and chewing gum are examples.

Severe trauma may cause serious hearing losses by rupture and/or dislocation of the eardrum and ossicles (Figure 3.1). The inner ear may be damaged by dislocation of the stirrup or fracture of the temporal bone containing the cochlea. A sudden intense pressure wave (such as produced by an explosion or a blow to the external ear) may cause hearing loss which may be partial or total, temporary or permanent. Not unusual is unintentional puncture of the eardrum as a result of attempting to remove wax from the canal.

Prolonged noise exposure can cause hearing loss. The intensity, noise frequency, duration of exposure, and age of the person must be considered before it can be proved that the hearing loss was caused by noise.

Rapid changes in altitude can damage the middle ear. This occurs most frequently during airplane trips in those suffering from colds, sinusitis, or other infections. The greatest difficulty arises during descent, when air cannot enter the Eustachian tube from the throat to equalize middle ear pressure.

Certain substances with potentially toxic side effects, such as quinine and its derivatives, nicotine, and salicylates, are possible causes of hearing loss in susceptible people. This is reversible if the drug is discontinued at the onset of ringing of the ears (tinnitus) or auditory nerve irritation. Continued use of the drug may kill the auditory nerve or cause disturbances of the labyrinth (difficulty with balance).

Several diseases (mumps, measles, scarlet fever, diphtheria, respiratory infections, etc.) have been responsible for hearing loss in adults. The effects may be due to direct infection in the middle and/or inner ear, or as a secondary infection in the middle ear.

This outline of causes of hearing loss is not intended to be exhaustive, but to describe the major reasons people lose hearing. To determine exactly what has caused a particular person to suffer a loss of hearing requires a medical history and an occupational history. A simple audiometric test can measure the hearing loss but cannot evaluate the cause. Loss of hearing is only a symptom of damage to the hearing mechanism; multiple causes may exist. To determine if the hearing loss is due to noise exposure or to other causes is sometimes difficult and usually requires an otologic consultation.

3.1.2 EFFECTS OF EXCESSIVE NOISE

Sound or noise of sufficient intensity and duration will produce loss of hearing. Sudden or blast-like noise on the order of 150 dB may cause permanent injury to the tympanic membrane, the ossicles, and the organ of Corti.

Noises in excess of 160 dB are beyond the anatomical safety factors of the ear and cause immediate damage. Less intense noises at about 130 dB may cause swelling of the hair cells of the organ of Corti on short exposure and destruction of these cells on longer exposure. These changes are usually localized rather than involving the entire organ of Corti. In man, sound levels of 120 dB produce marked discomfort, 130 dB a tickling sensation, and 140 dB, pain (Table 3.1). The destruction of the hair cells in the organ of Corti is an irreversible process and

TABLE 3.1 Sound Levels in Decibels for Construction and Other Environments[a]

Sound level (dB)	Activity
130	Near jet aircraft at takeoff
120	
110	
100	Typical (outdoor) construction noise (15 m)
80	Pile drivers, power lawn mower
60	Air compressors and concrete mixers
50	Conversation (at 1 m)
40	Average business office
20	Public library
0	Recording studio
	Threshold of hearing (1000 to 40,000 Hz)

[a] Reference sound pressure 2×10^{-5} N/m^2 corresponds to 0 dB sound level. An increase of 6 dB in sound level corresponds to a doubling of the sound pressure. Adapted from Olishitski and Harford, 1975.

the resultant hearing loss is permanent, whereas the initial swelling is reversible and causes only a temporary hearing impairment.

One serious problem requiring solution is the detection of noise-susceptible individuals so they may be protected from damaging noise exposure. Studies observing time of induction of auditory fatigue, time for recovery of auditory acuity after noise exposure, brain-nerve action potential, and so on, have given no practical solution to the problem. Before a practical tool can be devised, more observation and experience are required. To date, periodic audiometric examinations, along with workplace noise level measurements, offer the only practical approach to the problem.

The intensity of noise, the frequency, the length of exposure, and individual tolerance are the most important factors influencing loss of hearing. However, the age of the person, the presence of coexisting ear or upper respiratory tract disease, the influence of certain drugs or nutritional states, the distance from the source of noise, the position of the ear in relation to sound wave production, and the character of the exposure (continuous, intermittent, sporadic) are additional factors influencing auditory damage by noise.

It has been reported from Sweden that low frequency sounds (infra sounds) influence human performance in man at pressure levels below the threshold of hearing. Frequencies at levels below 20 Hz have resulted in increased reaction time and nausea, and at higher levels balance disturbances have been produced (Arbetarskyddsstyrelsen, 1978).

Hearing losses due to excessive noise exposure are usually about the same in both ears unless the working position is such that one ear is constantly in direct line with a single noise source. Total deafness rarely occurs from industrial exposures to continued noise of high level. (Sudden, severe concussion may cause immediate, irreparable, and complete hearing loss.)

A person with auditory nerve damage (from acoustic trauma or presbycusis) first loses hearing in the higher frequencies. In more severe cases, the ability to hear the higher frequencies of conversational speech may be impaired (at around 4000 Hz). A person thus afflicted may hear the lower range of a man's voice better than a woman's higher pitch and may not hear the telephone ring. He/she soon begins to speak louder and louder and in a monotone since the modulating effect of hearing is impaired. Because low tones are heard better than higher ones, noise seems unduly loud and conversation becomes difficult in a noisy environment. The person may have difficulty in telephone conversations; he/she hears people in group conversation but cannot understand what they are saying regardless of loudness of speech. Amplification of sound may not improve the situation; in fact it may add confusion due to distortion of sound.

In contrast to this individual, one with middle ear or conductive deafness (the ossicles of the middle ear fail to conduct accurately the frequency and amplitude of sound to the inner ear) will complain that others in conversation do not speak loudly enough. Understanding is not impaired if the sound level is sufficiently high; this person can usually hear adequately in noisy places and has little trouble with telephone conversation. As the conduction defect worsens, difficulties in communication may be encountered.

Aside from hearing loss and physical damage to the ear, another aspect of the problem of noise is the psychological effects of working in noisy surroundings. Some studies have indicated that when noise levels are reduced, work output increases, fewer errors are made, and employee satisfaction increases. There is also evidence that chronic exposure to high noise levels causes physiological changes such as irritability, fatigue, and increased blood pressure and heart rate.

Measurements of hearing are referred to the "zero reference level" for normal hearing (see Figure 3.2). These measurements are expressed in terms of decibels of hearing loss compared to average normal hearing. A "hearing" score of zero indicates no loss compared to the average; increasing scores indicate hearing loss in decibels.

The first and most notable damage caused by excessive noise is a hearing loss at about 4000 Hz. However, there is extreme variability in individual reactions to noise. Similar loss of hearing may occur because of aging (presbycusis). It is therefore difficult to determine the causes in some individuals where age may be a factor.

Figure 3.2 shows two audiograms. The top audiogram illustrates a slight hearing loss due to industrial noise and the bottom one shows a severe hearing loss.

3.1.3 MEASUREMENT OF SOUND

Sound level meters are used for the objective measurement of sound. They consist of a microphone, an amplifier, and a meter that gives a visible reading on a scale. Most meters incorporate three different types of weighting of the sound. These are known as the A, B, and C scales. These are designed to simulate sound hearing sensitivity at different frequencies for different sound levels. Figure 3.3 gives the weighting characteristics of the filters. Particularly the dBA scale has achieved widespread use for many practical applications. It can give a fairly

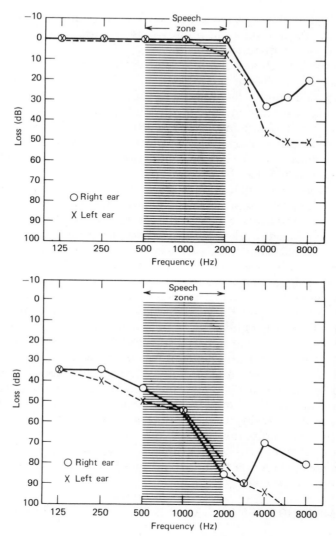

Figure 3.2 Hearing losses in decibels for various frequencies for a slight hearing loss (top) and for a severe hearing loss (bottom). ○ Right ear; × left ear.

good prediction of the subjective response to noise and can be used to assess the risk of hearing damage. In other words, the dBA scale simulates the ear.

The dBA scale is referenced to a sound pressure level equal to 0.00002 N/m^2, which corresponds to the threshold of hearing. To calculate the sound pressure level in decibels, the following formula can be used:

$$L_P = 20 \log \frac{P}{P_0} \, \text{dB}$$

where P = measured rms sound pressure
P_0 = reference sound pressure (0.00002 N/m^2)

From the formula it can be derived that a doubling of sound pressure corresponds to an increase of 6 dB.

3.1.4 FEDERAL REGULATIONS FOR NOISE EXPOSURE

Federal regulations permit the worker to be exposed to a noise level of at most 90 dBA for 8 hr/day and prescribe a reduction of the permissible exposure time by 50% for each 5 dBA increase in noise level (see Table 3.2). For a worker who is exposed to different noise levels throughout the day, the total permissible noise dose can be obtained

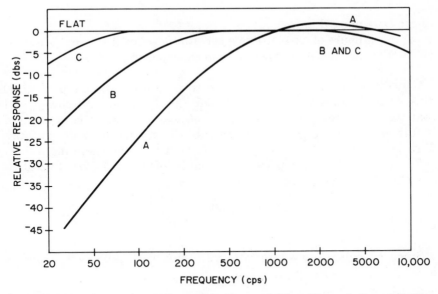

Figure 3.3 There are three different frequency weighting filters, A, B, and C. They modify the dB readings and produce an overall weighted average. The dBA filter is the most frequently used, since it simulates the sensitivity of the human ear. The filters can be attached to a sound level meter.

TABLE 3.2   Occupational Safety and
Health Act of 1970, Permissible Noise
Exposures

Duration per day (hr)	Sound level, slow response (dBA)
8	90
6	92
4	95
3	97
2	100
$1\frac{1}{2}$	102
1	105
$\frac{1}{2}$	110
$\frac{1}{4}$ or less	115

by adding all the noise doses corresponding to the various noise levels according to the following formula:

$$D = \frac{C}{T}$$

where D = noise dose
C = hours of exposure to a certain noise level
T = permissible duration of noise exposure

Consider, for example, a machine that subjects its operator to 90 dBA while it idles and to 95 dBA while it is used at full power, and that the machine operates at idle for 30% of the time it is in use and in full power 70% of the time. From Table 3.2 the permissible exposure duration for 90 dBA is 8 hr and for 95 dBA is 4 hr. Assuming a total of 7 hr of use per day, implying 2.1 hr at 90 dBA and 4.5 hr at 95 dBA, one obtains a total noise dose of

$$\frac{2.10}{8} + \frac{4.90}{4} = 1.487$$

Since the noise dose is larger than 1.0, this work exposes its operator to a noise that is not permissible.

3.1.5   ASSESSMENT OF THE EFFECTS OF NOISE ON WORKERS

Medical supervision of workers exposed to noise is essential for the benefit of employee and employer. The objective is to make sure that no damage is done to the hearing. This can be accomplished by several means: examination, proper job placement, and protection against noise. Prior to the preemployment physical exam, it is advisable to obtain a detailed history from the applicant covering prior occupational experience and a personal record of illness and injuries. This is important since most physicians feel that a carefully taken history is indispensable for good diagnosis.

For noisy work environments, the history should include information on noise exposures in previous jobs as well as in military service and should embrace occurrence of earache, ear discharge, ear injury, surgery (ear or mastoid), head injury with unconsciousness, ringing in the ears, hearing loss in the immediate family, the use of drugs, and history of allergy and toxic exposures. Permanent records should be made of each examination.

3.1.6 AUDIOMETRIC EXAMINATION

The examination of the hearing mechanism is supplemented by tests of hearing acuity. These are made using an audiometer. Audiometry is an essential part of any hearing conservation program. It facilitates the detection of changes in auditory acuity since it enables hearing losses at 4000 Hz to be detected long before the victim notices any difficulty in conversation. An audiometer can also be used to evaluate the effectiveness of ear protectors and noise control measures. Other types of tests such as the spoken voice, whisper, and watch-tick tests are not sufficiently accurate to be of any use.

Pure tone audiometry is the recognized method of testing hearing ability. An audiogram is obtained by presenting pure tones of variable levels of intensity until each tone is just heard. This is the threshold for each tone. Each ear is tested separately. Recommended test frequencies are 500, 1000, 2000, 3000, 4000, and 6000 Hz. Hearing losses less than 15 dB are considered normal and do not require further examination. Losses in excess of 20 dB indicate the need for further study.

The best measurement of hearing is obtained if tests are made before noise exposure, such as at the start of the shift or after a weekend away from work. This is because of the condition known as temporary threshold shift, which means that the hearing is poorer after exposure to excessive noise and recovers after a period of quiet.

Audiometric tests must be done in an approved chamber, called an audiometric booth and be located in an area that is as quiet as possible. These chambers are commercially available. The occupational audiometric technician need not be an audiologist; nurses or technicians with special training may properly give audiometric tests. Diagnosis and determination of the cause of hearing loss, however, are medical functions and may require the assistance of a consulting physician or otologist.

Management sometimes claims that audiometric schemes are too expensive. But the cost of a single successful compensation case may considerably exceed that of the necessary testing equipment. Experience has shown that annual audiometric tests are sufficient. In plants with continuous medical controls, tests may be conducted at longer time intervals, which should not exceed 2 years if the workplace noise is more than 85 dB.

Proper job placement of workers, that is, matching their physical capacities to the physical demands and exposures of the job, is an established function of the occupational physician. Exposure of workers to noise involves application of the same principles. A worker with a confirmed hypersensitivity to noise should not be so exposed, whereas the otherwise qualified, totally deaf individual would be ideal because of the absence of any possibility of further damage. An individual with a partial hearing loss due to exposure to excessive noise should be protected from further loss by proper job assignment, by the use of adequate engineering controls, or by the use of personal protective devices.

The ideal method of preventing noise damage to the ear is by control or reduction of the noise at its source by engineering methods. When these controls are not feasible or are only partially successful, it may be necessary to resort to protection of the workers through the use of earplugs, earmuffs, or other suitable devices.

3.1.7 HEARING PROTECTORS

There are two types of hearing protectors that can be used by construction workers, earplugs and earmuffs. The plugs are designed to occlude the ear canal. Many types of materials are available. Cotton has traditionally been used but, unfortunately and contrary to popular belief, it affords no protection!

The ready-made plugs are made of rubber, neoprene, or plastic materials. They are designed to fit the ear canal and to prevent hearing loss by reducing the level of noise reaching the eardrum. Recently, custom molded earplugs have been made available. These are individually made to fit the external part of the ear, protruding slightly into the ear canal. Once made correctly, these can offer excellent fitting and protection and have a long life.

Earmuffs, sometimes called over-the-ear cushion-type protectors, are designed to cover the entire external ear. They consist of ear cushions made of soft spongy material or specially filled pads to assure snug and comfortable fit. The muffs are held firmly in place either by adjustable metal headbands or by appropriately designed caps or hoods.

It has been found that employees are more willing to wear ear protectors when they have been given a choice in selecting the type and size best suited to them. It is advisable, therefore, to have available several types of protectors and to explain the various features.

The sound attenuation provided by ear plugs varies between 15 dB for low-frequency sounds to 35 dB or more for high frequencies. At frequencies above 1000 Hz, muffs provide about the same protection as plugs. At frequencies below 1000 Hz, certain muffs provide more protection than plugs.

Earplugs and earmuffs may be worn together in intense noise. This combination provides an extra attenuation of approximately 5 dB.

Supervisory personnel should inform their employees who work in excessive noise of the benefits to be derived from hearing conservation. As far as the new worker is concerned, the logical time to acquaint him with hearing protection is at the very beginning when he is also being briefed on first-aid facilities, personnel protective equipment, and the like. It should be explained that the wearing of ear protectors in noisy situations is as necessary for individual protection as are safety goggles, safety shoes, hard hats, and respirators. The development of an ear protector program is therefore no different from any other protective program designed to eliminate or reduce an industrial hazard. Ear protectors should be available to all employees who need them, and their use should be mandatory.

Workers who regularly wear ear protection report that they actually hear better in excessively noisy areas. Cutting down the noise level reaching the ear decreases the distortion in the ear so that speech and warning signals are actually heard more clearly. An analogy can be drawn with wearing sunglasses to reduce excess glare and thus improving vision.

Regular inspection of ear protectors is necessary because of deterioration with usage and time. Workers should discard protectors as soon as any defect is noted.

3.2 OCCUPATIONAL EXPOSURE TO VIBRATION

In today's occupational environment, mechanical vibration is frequently encountered. High exposure to vibration presents a health hazard and extreme levels of vibration may result in tissue damage.

There are two major kinds of vibration: *whole body vibration* and *segmental vibration*. Common sources of whole body vibration are vehicles of all types, such as long-haul or over-the-road trucks and related vehicles, buses, earthmoving equipment, forklift trucks, and other industrial moving machines.

Segmental vibration, or vibration of the extremities, is often induced by hand-held tools. Among those tools are pneumatic devices such as jackhammers or concrete-breaking devices, hand-held oscillatory, rotary tools such as those used for vibration of concrete, and power drills, saws, and tampers. Gasoline-powered chain saws have also been widely implicated in vibration exposure, creating some degree of distress, discomfort, as well as reports of certain physiological and pathological changes in man.

3.2.1 WHOLE BODY VIBRATION

A major source of whole body vibration in the construction industry is transportation vehicles. The vibrations may be of a single frequency but they are usually complex and random and often found as a combination of different frequencies. Figure 3.4 shows a seated driver. This figure emphasizes that different parts of the body have different resonance frequencies, for instance, 3 to 5 Hz for the shoulder and the stomach. This might very well be the reason why this frequency range produces the largest discomfort. Laboratory studies have shown that vibration between 3 and 5 Hz is likely to be physically uncomfortable at acceleration levels of approximately 0.1 g; painful and distressing

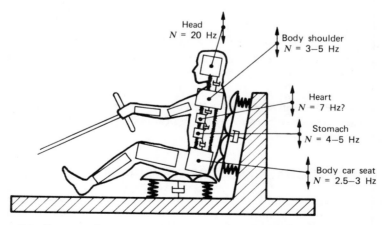

Figure 3.4   Resonant frequencies for different parts of the body for a seated operator.

at intensities in the region of 1 g; and injurious if the acceleration exceeds 2 g and is sustained for more than a few cycles of motion. These findings form the main background for the present ISO (International Standards Organization) standards for vibration (see Figure 3.5). (Mackie et al., 1974; Miwa et al., 1975 Gruber, 1976).

However, there are also psychological factors that can greatly influence the discomfort of vibrations. These factors are as follows:

1. *The Nature of the Task.*   For example, riding a recreational boat is usually associated with pleasure, although the same magnitude of vibrations would be perceived as stressful in most working environments.

2. *The Duration of the Task.*

3. *The Person's Degree of Training or Familiarity with the Task.* For example, a skilled horseback rider can compensate for much of the vibrations by tensing specific muscles.

4. *The Presence of Other Stressors Acting in Combination* (for example, noise).

In addition to discomfort effects, there are several reputed health effects such as various spinal, anal–rectal, and gastrointestinal disorders. Exposure to vibration also influences some physiological responses. (Fothergill and Griffin, 1977; LaDou; 1980).

The most common physiological responses to tolerable levels of vibrations have been reported to be an increased heart rate, with 10 to 15 beats/min above the resting levels, and promptly returning to normal after vibration ceases; blood pressure increases at frequencies around 5 Hz, but blood pressure is found to decrease slightly in the 10 to 20 Hz range. Some investigations have revealed mild hyperventilation and slight respiratory rate increases as a function of intensity, and oxygen uptake increases significantly in the frequency range of 5 to 7 Hz. One of the most notable findings is that the visual acuity decreases, particularly for vibrations of 10 to 25 Hz, which is thought to represent the resonant frequency of the eyes. As a result, there is often a reduction in the operator's performance level (Collins, 1973; Grether, 1971).

Hansson et al. (1976) studied exposure to whole body vibrations of drivers of 44 industrial trucks and 14 mining industry machines. They investigated boring and drilling machines, giant trucks (Euclid type), various forklifts, large excavating shovels, front-end loaders, graders, and other machines. Their studies indicated that vibration in drills was slight. However, there was severe vibration in wheel-loaders in the mining industry. Vibration in three of the six machines presented a risk of health during 8 hr of exposure (according to the ISO standards). Vibration presented a risk to health in 6 of the 41 industrial trucks studied if exposure lasted 8 hr. Vibration was fatiguing and reduced the work capacity of the driver in two-thirds of the machines studied (according to ISO standards). In the industrial trucks, vertical resonance was recorded in the 2 to 10 Hz range with resonance frequency depending on the size of the truck. The large and heavy trucks displayed a lower frequency than the small, light trucks. Drills differed from the other machines by having a high resonance frequency. The other machines studied had resonance frequencies under 10 Hz.

3.2.2   SEGMENTAL VIBRATION

Within the scope of this text, the effects of localized vibration from the use of hand-held tools and equipment, transmissible through the var-

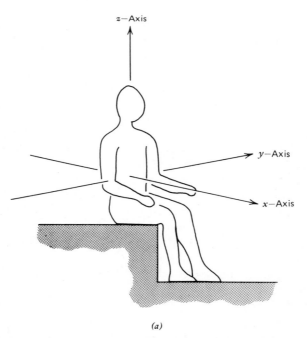

(a)

Figure 3.5 (a) Definition of coordinate system for measurement of vibration. (b) ISO limits for longitudinal (z-axis) and transverse (x- and y-axes) whole body vibration. Limits for maximum exposure (——), fatigue-decreased proficiency (------), and reduced comfort (—·—·—) are given for 24-hr exposure and 1-min exposure.

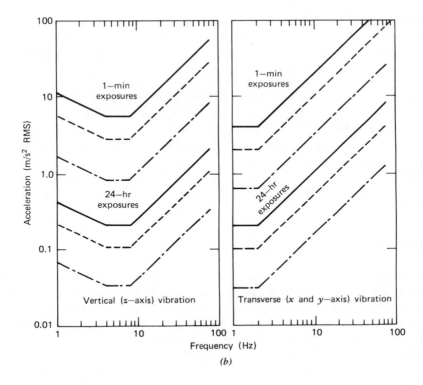

(b)

ious body members, is of greatest importance. The tools mentioned earlier which are operated by air and other hand-held impact tools, especially where there is much oscillation or some imbalance of the equipment, may produce considerable localized phenomena. It has been well recognized for nearly a century that a condition known as Raynaud's disease (or white finger disease)can occur from extensive and prolonged vibration. Raynaud's disease is produced by frequencies between 50 and 100 Hz, such as in pneumatic drills. This entity consists of stiffness and/or numbness of the fingers or hand, a gradual loss of muscle control such as inability to hold, grasp, or manipulate, finger swelling, numbness and tingling, and, in advanced cases, blanching of the fingers along with bluish discoloration. Raynaud's phenomenon is greatly aggravated by low temperatures and dampness as well as by smoking. Symptoms and signs may be of varying duration of several hours or persisting for days. If diagnosed early and preventive techniques are instituted, this phenomenon is reversible without continuing effects. Most often, the vibrating stress must be removed or decreased or the worker transferred to other work. If allowed to continue, this condition may become irreversible. With continuous vibration stress, particularly low frequencies and high intensities as may be seen with some heavy impact tools, there can be additional, localized damage to the bones of the hands as well as to the joints of the fingers, hands, and wrists. In advanced cases, actual decalcification of some areas of the bones of the hands, wrist, and forearms may be observed by X-ray. There may be a resultant restriction of motion, along with painful joints.

Another syndrome less common than Raynaud's disease is Dart's disease. In Dart's disease, the hands become blue, painful, and swollen (Allen, 1971).

3.2.3 PREVENTION OF VIBRATORY PROBLEMS

Reduction of vibrations through engineering controls, isolation of the worker from the vibratory forces, and a reduction of the duration of exposure to vibration are key factors. As with conservation to prevent occupational hearing loss, vibration may be controlled and reduced at its source by appropriate engineering techniques such as the use of vibration isolators; proper machine maintenance such as balancing and replacement of worn parts; substitution of equipment and processes that produce high levels of vibration for those producing less vibration; and modification of speeds and pressures needed for operating equipment as well as ascertaining the worker's physical responses and sensitivities to these stresses. Within vehicle driving, attempts are made to use damping devices on the driver's seat. Typically, these are cushions with springs or small hydraulic devices intended to deflect as much body movement as possible. These modifications are difficult to make. From a practical viewpoint, a human's legs are generally superior to most seating devices as they have many automatic compensating postural and muscular changes to minimize vibration energy being received.

3.3 OCCUPATIONAL EXPOSURE TO ELECTROMAGNETIC RADIATION

From a physical point of view, light is radiant, electromagnetic energy that can be seen. Visible light covers only a narrow portion of the entire electromagnetic spectrum with wavelengths from 10^{-4} to 10^{-5} cm. The entire electromagnetic spectrum includes wavelengths from about 10^7 to 10^{-12} cm (see Figure 3.6).

This tremendous range includes cosmic rays, X-rays, ultraviolet rays, the visible spectrum, infrared rays, radar, FM, TV, and radio broadcast waves. In this section we discuss some of the health hazards associated with electromagnetic radiation, such as infrared radiation, ultraviolet radiation, microwaves, and lasers.

3.3.1 MICROWAVES

Wavelengths between 10^2 and 10^{-2} cm are referred to as microwaves (see Figure 3.6). The large wavelengths contain the ultrahigh-frequency radio waves; the middle portion is used in radar. The small wavelengths are the ones used for cooking and drying.

Technology using microwaves has been developed in the past few years. Microwave systems have become so inexpensive that they are frequently used in home cooking. But the greatest use is in industry for such purposes as drying match heads, veneers, laminates, and paper; for molding and curing plastics; and in the food industry for drying, thawing, and precooking. Microwave generators and radar are also commonly used for military purposes, such as communications and navigation.

The microwaves that are closest to the infrared portion of the spectrum produce heat but no penetration. They can cause damage to surface tissue, for example, to the cornea of the eye. Wavelengths greater than 3000 MHz (megahertz) can penetrate the skin and cause heating within the tissue; in the eye, the lens will be damaged, as may the iris. Such damage can occur very fast, and there is a danger that the exposed person might not receive any prewarning (since there are no receptors for heat inside the body). Based on available evidence, lens damage can probably not occur in a human without associated severe facial burns (Appleton, 1973; Appleton, 1974; Mild et al., 1979).

Before personnel are assigned to work in or around radar equipment, they should have medical examinations, including a complete eye study. It should also be noted if there are any metal implants in the body. Personnel working with or near microwaves should receive periodic physical examinations (Eure et al., 1972; Hathaway et al., 1977; Mild et al., 1979).

The maximum exposure allowed is called the "threshold limit value" (TLV). For frequencies between 300 MHz and 300 GHz (300,000 MHz), the maximum dose (TLV) is 10 mW/cm^2 for a period of 0.1 hr.

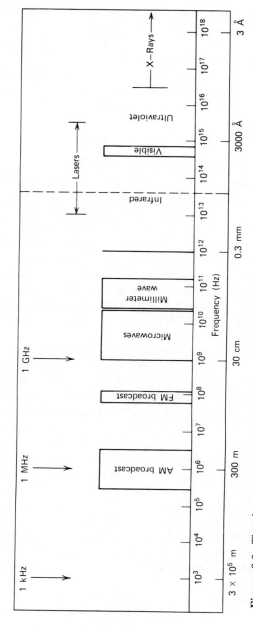

Figure 3.6 The electromagnetic spectrum. The wavelength in meters is obtained from wavelength (m) = $3 \cdot 10^8$/frequency (Hz), where $3 \cdot 10^8$ is the velocity of electromagnetic radiation (e.g., light) in air.

3.3.2 ULTRAVIOLET RADIATION

People who work outdoors, exposed to sunlight, are chronically exposed to ultraviolet (UV) radiation. The chief area of concern with regard to ultraviolet radiation is the possibility of damage to the eyes and skin irritation. The most common reaction to ultraviolet exposure is sunburn, appearing as simple redness of the skin (erythema). In more severe cases, the skin will get burnt sufficiently to cause discomfort, pain, blistering, and sometimes incapacitation.

Bioeffects studies, including epidemiological studies of human populations, also indicate possible long-term effects of ultraviolet radiation such as premature skin aging and malignant skin tumors as a result of prolonged or repeated exposure.

Indoor jobs, especially welding, can also produce damage related to ultraviolet exposure. Other, nonnatural, high-intensity ultraviolet sources, such as intense lights, may produce skin irritation in a matter of seconds, depending on distance. Photokeratitis of the skin has been characterized by a period of latency, tending to vary inversely with severity of exposure. The latent period may be as short as 30 min or as long as 24 hr, but is usually around 6 to 12 hr long. This condition is one of thickening, hardening, or loss of skin elasticity, along with hyperpigmentation (a darkening) of the skin. Erythema of the skin of the face, especially that surrounding the eyelids, commonly occurs, as well as conjunctivitis from prolonged exposure to ultraviolet.

The International Commission on Illumination has divided the ultraviolet spectrum into three different wavelength bands: 200 to 280, 280 to 315, and 315 to 340 nm. Wavelengths below 200 nm are of little biological significance. Permissible exposure limits have been established by the National Institute of Occupational Safety and Health. These emphasize that the greatest danger of eye and skin injury is produced by the electromagnetic radiation above 315 nm, with the maximal allowable limit of 1 mW of ultraviolet radiation per square centimeter for exposures exceeding 1000 sec (U.S. Dept. of Health, Education and Welfare, 1977). This document also describes technical problems in measuring UV energy and recommendations for protective clothing, work practices, medical examinations, and record-keeping information (Ramsey, 1975).

Perhaps the most common source of nonsolar UV radiation is that produced by welding. Skin and eye problems that occur from exposure to the UV resulting from welding are identical to those occurring from sunlight. The hazards from welding depend on the distance of the worker from the arc, the angle at which the rays strike the worker's eye, the radiation intensity, and the type of eye protection worn by the worker. There are also indirect sources of radiation such as produced by reflection from metal surfaces, walls, and ceilings. Metal surfaces, especially aluminum and stainless steel, can reflect enough UV radiation to pose a danger to nearby workers. Installation of shielding around welding areas to protect other workers is a good practice (Emett and Horstman, 1976).

Other important, but indirect, effects of UV radiation may be produced at work. It has long been known that ozone (O_3) may be produced in considerable quantities from a welding arc by a chemical reaction of the atmospheric oxygen with ultra energy less than 200 nm. Ozone is a powerful oxidizing compound with an exposure threshold limit value (TLV) of 0.1 ppm or 0.2 mg/m^3, and is considered highly toxic. It produces irritation of the eyes and upper respiratory system. On prolonged exposure, ozone will produce respiratory distress such as shortness of breath, coughing, and pulmonary edema. Investigations have shown that more ozone is produced during the welding of aluminum than during the welding of steel. Likewise, more ozone is produced when argon gas is used for shielding than when carbon dioxide is used (TLV's: 1979; The American Conference of Governmental Industrial Hygienists, 1976).

Ultraviolet radiation can also cause decomposition of some cleaners and degreasing agents such as trichloroethylene. The major products of such reactions are phosgene and dichloroacetyl. Phosgene is a toxic and rapidly acting gas, producing dermatitis on skin contact. Inhalation at higher concentrations can produce pulmonary edema, and extreme exposure leads to death from respiratory and/or cardiac arrest. The established TLV for phosgene is 0.05 ppm, but the toxicity of dichloroacetyl is not as well known and no threshold has been established for it. Studies have shown that using gas metal-arc welding in areas where trichloroethylene is used produces a phosgene concentration of 3 ppm, well above the TLV. Welding areas therefore should not be located near metal cleaning or degreasing work. Special precautions must also be taken when welding metal parts that have been degreased in trichloroethylene or other chlorinated hydrocarbon agents.

Another important source of UV radiation is mercury vapor lamps. There are presently over 25 million high-intensity mercury vapor discharge lamps used in the United States. The FDA's Bureau of Radiological Health reviewed reports of more than 460 injuries in 1979 involving these lamps (U.S. Dept. of HEW, 1977). Most of the patients in these cases complained about eye discomfort and skin burns. Injuries occur when the protective outer glass bulb of these lamps is broken, permitting an exposure to short-wave UV radiation. Most such injuries occur following continuous exposure for several hours at a distance of up to 6 m from the damaged lamp. These lamps can continue to operate and emit hazardous UV radiation for many days.

The agency recommends that individuals exposed to UV radiation from a damaged lamp should see a physician if symptoms of skin burns or eye irritation occur. Individuals using mercury vapor or metal halide lamps in a home or yard should consider replacing them with those that have a safety shutoff feature. An alternative is to install a UV radiation-absorbing shield over the lamp reflector.

There are several other types of illumination that emit UV radiation. High-pressure xenon arcs emit a continuous spectrum similar to that of sunlight. The newer "plasma torches" may produce temperatures in excess of 6000°C, close to the temperature of the surface of the sun, and emit intense ultraviolet radiation. Other incandescent sources emit little UV energy except at temperatures above 2300°C.

3.3.3 INFRARED RADIATION

Excessive exposure of the eyes to luminous radiation, mainly visible and infrared (IR) radiation, from furnaces and similar hot bodies has been reported for many years to produce "glassblower's cataract" or "heat cataract," the development of an opacity of the rear surface of the lens of the eye.

Infrared light (or heat) extends from the visible red light region (750 nm) to the 0.3 cm wavelength of microwaves (see Figure 3.6). Infrared radiation is transmitted from any surface that has a higher temperature than the surrounding environment.

Exposure to infrared radiation increases the temperature of the skin. The amount of increase depends on the wavelength and the exposure dose. Exposure to IR radiation in the region between 750 and 15,000 nm can cause acute skin burns and increased and persistent skin pigmentation. This short wavelength region of the infrared can cause injuries to the cornea, iris, retina, and lens of the eye, the main organ of concern in man. IR radiation in the far wavelength region of 5000 nm to 0.3 cm is completely absorbed in the surface layers of the skin.

Infrared radiation is used in a large number of industrial processes, such as drying and baking of paints, varnishes, enamels, adhesives, printer's ink, and other protective coatings; heating of metal parts for shrink fit assembly, thermal aging, brazing, radiation testing, and preparing surfaces for application of adhesives and welding; dehydrating of textiles, paper, leather, meat, vegetables, potteryware, and sand molds; and spot and localized heating.

The available data indicate that acute ocular damage from IR radiation can occur with energy densities between 4 and 8 W·sec/cm^2 incidence upon the cornea, depending on the wavelengths absorbed. It is generally believed that a maximum permissible dose of 0.04 to 0.08 W·sec/cm^2 would prevent any acute effects as well as the more chronic effects on the eye (Ketchen et al., 1978).

Special protective eyewear and face shields are available for protecting workers to these radiations.

3.3.4 LASERS

The light from a conventional light source radiates in all directions. Light of varying wavelengths and phases can amplify or extinguish. To obtain a directional beam from a conventional source of light, it must be focused, and even then it soon spreads its beam again. Light from a laser, however, vibrates in a single plane, travels in only one direction, and is monochromatic (of the same wavelength). Light such as this is known as coherent light. Nearly all lasers are potential eye hazards. All produce extremely high-intensity light radiation of a single wavelength (or a narrow band of wavelengths), depending on the material used for light amplification (Sliney, 1968).

Although a laser produces only one wavelength or frequency, laser units can be designed for a wide range of frequencies. Laser beams are not limited to the frequency of visible light. One of the most powerful beams uses a carbon dioxide laser and shoots out a continuous beam of intensely hot but invisible infrared radiation. Some other lasers operate in the ultraviolet frequencies.

There are three types of laser-generating media: (1) the solid state, of which the ruby crystal is the most common type; (2) the gaseous state, of which the helium–neon is the most common; and (3) injection or semiconductor.

The most common type of laser uses a synthetic ruby crystal made from aluminum oxide and chromium. The ruby crystal has the shape of a rod and is surrounded by a flashtube. One end of the ruby rod is silvered to reflect light; the other end is partly silvered. When the flashtube flashes a burst of light, it excites some of the chromium atoms, raising them to a higher, unstable energy level. The excited atoms drop back to their stable energy level, giving off photons of light. The photons oscillate between the reflecting ends of the ruby rod, exciting other chromium atoms, which produce more photons. As a result a stream of photons will burst through the partly silvered end of the rod in one swift pulse of coherent light. The entire sequence takes only a few thousandths of a second.

The gaseous laser uses a similar principle but a different energy source. In a helium–neon gas laser, a mixture of helium and neon is required to provide an active medium. A radio frequency or direct current excites the helium atoms. The helium atoms excite the neon atoms to a higher energy level, and the neon atoms return to the normal state, giving off photons. the output beam is increased by repeated passes between the reflecting end mirrors, triggering more photons. Gaseous lasers produce a continuous stream of light known as a "continuous wave." The maximum laser power is lower than that produced in solid state pulse lasers.

The injection laser uses a tiny semiconductor crystal with the atoms arranged in a lattice structure. A very small quantity of light will cause emission of photons. Injection lasers can approach power levels that could be dangerous.

The laser beam travels in parallel lines and does not spread out as ordinary light does. The energy of the laser is, therefore, confined to a small diameter. For example, an 18-W argon gas laser can generate power 2000 times as strong as sunlight. In contrast, the light from a 20-mW helium–neon laser would not be felt on the skin unless the beam is focused to a very small area. If the beam were focused to an area of 1 mm^2, the power density would be 2 W/cm^2, which is a potential hazard.

A ruby laser generally emits energy in short bursts. The energy usually is 200 to 500 J in pulse widths of 175 to 350 msec. These types of lasers are extremely strong. For example, a 500-J laser will produce 1400 W in a 350 msec burst. If this beam is focused to 1 mm^2, it can pierce a diamond in a fraction of a second.

The type of laser that has found the greatest use in the construction industry is the helium–neon (HeNe) gas laser. Its highly collimated beam is used to project a reference line for construction equipment in such operations as dredging, tunneling, pipe laying, bridge building, and marine construction. It is especially useful in operations over large bodies of water where height references are difficult to establish. Several manufacturers provide small helium–neon lasers complete with transit mounts and collimating optics for this purpose. These lasers generally have a power output of 1 to 10 mW. Most small helium–neon lasers have a beam diameter of 1 to 3 mm which is expanded by collimating optics to 20 to 30 mm (approximately 1 in.), thereby making it nonhazardous. Some collimation systems are designed to provide a fan-shaped beam so that a reference plane is produced rather than a line. In other applications, the beam is directed at another reference point, such as a target card. A small helium–neon laser is also widely used for precise distance measurements in surveying.

Outside the construction industry, lasers have been used for welding and micromachining fine parts. A laser photocoagulator is used by surgeons to repair torn retinas. The laser makes bloodless surgery possible. A laser can also be used to destroy malignant skin tissue, remove birthmarks, or burn away warts.

Severe damage to the eye has been reported from accidental exposures to laser radiation. It is therefore very important that workers using laser equipment be properly instructed in the safe use of that equipment. It is especially dangerous to look directly into a laser beam without eye protection. A laser beam can damage either the cornea or the retina, depending on the wavelength. It should be remembered that smooth and reflective surfaces can redirect a laser beam and cause eye damage to workers not looking directly into the beam. Such items as glass-covered pictures and polished door handles should be removed from areas where lasers are used (Sliney, 1977).

The American National Standards Institute has published safety rules for lasers. These provide technical information on measurement, calculation, biological effects, criteria for exposure of eyes and skin, hazards of explosion, and the administration of safety programs (American National Standards Institute, 1973).

REFERENCES

Allen, G. Human Reaction to vibration. *Journal of Environmental Science*, 10–15, September–October 1971.

American Conference of Governmental Industrial Hygienists. *A Guide for Control of Laser Hazards*. Cincinnati, Ohio: 1976.

American National Standards Institute. *ANSI Z136.1*, New York: 1973.

Andrews, H. L. *Radiation Biophysics*. Englewood Cliffs, N.J.: Prentice-Hall, 1974, pp. 170–281.

Appleton, B. *Results of Clinical Surveys of Microwave Ocular Effects* (U.S. DHEW Publ. (FDA)73–8031). Washington, D.C.: U.S. Government Printing Office, February 1973.

Appleton, B. Microwave Cataracts. *Journal of the American Medical Association*, **229**(4) 407, 1974.

Arbetarskyddsstyrelsen, *Infra and Ultra Sound* (Directions No. 110:1). Stockholm: May 1978.

Collins, A. M. Decrements in Tracking and Visual Performance During Vibration. *Human Factors*, **15**, 379–393, 1973.

Emmett, E. A. and Horstman, S. W. Factors Influencing the Output of Ultraviolet Radiations During Welding. *Journal of Occupational Medicine*, 18, 41, 1976.

Eure, J. A., Nicholls, J. W., and Elder, R. L. Radiation Exposure from Industrial Microwave Application. *American Journal of Public Health*, 62, 1972.

Fothergill, L. C. and Griffin, M. J. The Evaluation of Discomfort Produced by Multiple Frequency Whole-body Vibration. *Ergonomics*, 20, 263–270, 1977.

Grether, W. F. Vibration and Human Performance. *Human Factors*, 13, 203–216, 1971.

Gruber, G. J. *Relationships Between Wholebody Vibration and Morbidity Patterns Among Interstate Truck Drivers*. San Antonio, Texas: Southwest Research Institute, 1976.

Hansson, J. E., Klussel, L., Svensson, G., and Wikström, B. O. Working Environment for Truck Drivers—an Ergonomic and Hygienic Study. *Arbete och Hälsa*, 6, Arbetarskyddsstyrelsen, Sweden: 1976.

Hathaway, J. A., Stern, N., Soles, E. M., and Leighton, E. Ocular Medical Surveillance on Microwave and Laser Workers. *Journal of Occupational Medicine*, 19, 1977.

Ketchen, E. D., Porter, W. E., and Bolton, N. E. The Biologic Effects of Magnetic Fields on Man. *American Industrial Hygiene Association Journal*, 39, 1978.

LaDou, J. Health and Safety Hazards of Truck Drivers. In C. Zenz (Ed.), *Developments in Occupational Medicine*. Chicago: Year Book Medical Publishers, 1980.

Mackie, R. R., O'Hanlon, J. F., and McCauley, M. E. *A Study of Heat, Noise, and Vibration in Relation to Driver Performance and Physiological Status* (Human Factors Research, Inc. Tech. Rep. 1735). Goleta, Calif.: 1974.

Marshall, W. J., et al. *Evaluation of the Potential Retinal Hazards from Optical Radiation Generated by Electric Welding and Cutting Arcs* (Nonionizing Radiation Protection Special Study No. 42–0312–77). Aberdeen, Md.: U.S. Army Environmental Hygiene Agency, 1977.

Mild, K. H., Landström, U., and Nordström, B. Biological Effects of Electromagnetic Fields of Radiofrequency and Microwaves—Hazards and Norms. *Arbete och Hälsa*, 30, Arbetarskyddsstyrelsen, Sweden: 1979.

Miwa, T., Yonekawa, Y., and Kojima-Sudo, A. Measurement and Evaluation of Environmental Vibrations. *Industrial Health*, 11, 159–164, 1975.

Moss, C. E., et al. An Electromagnetic Radiation Survey of Selected Video Display Terminals. Cincinnati, Ohio: U.S. Dept. of Health, Education & Welfare, National Institute for Occupational Safety and Health, 1978.

Odland, L. T. Observations on Microwave Hazards to USAF Personnel. *Journal of Occupational Medicine*, 14(7), 1972.

Olishifski, J. B. and Harford E. R. (Ed.). Industrial Noise and Hearing Conservation. Chicago: National Safety Council, 1975.

Olishifski, J. B. (Ed.). *Fundamentals of Industrial Hygiene*, 2nd ed. Chicago: National Safety Council, 1979.

Powell, C. H. and Hosey, A. D. (Eds.). *The Industrial Environment . . . Its Evaluation and Control* (U.S. DHEW Public Health Serv. Publ. No. 614). Washington, D.C.: 1965.

Ramsey, J. D. Occupational Vibration, In C. Zenz (Ed.), *Occupational Medicine, Principles and Practical Applications*. Chicago: Year Book Medical Publishers, 1975.

Sliney, D. H. The Amazing Laser. *National Safety Congress Transactions*, 8, 38–42, 1968.

Sliney, D. H. and Franks, J. K. Safety Rules and Recommendation. *Laser Focus Buyers' Guide*, January 1977.

TLV's: *Threshold Limit Values for Chemical and Physical Agents in the Workroom Environment with Indented Changes for 1979*. Cincinnati, Ohio: The American Conference of Governmental Industrial Hygienists, 1979.

U.S. Dept. of Health, Education & Welfare, National Institute of Occupational Safety and Health. *Criteria for a Recommended Standard. Occupational Exposure to Ultraviolet Radiations*. Cincinnati, Ohio: 1977.

Van Pelt, W. F., Payne, W. R., and Peterson, R. W. *A Review of Selected Bioeffects Thresholds for Various Spectral Ranges of Light* (BRH Bull.). Rockville, Md.: U.S. Dept. of Health, Education & Welfare, Food & Drug Administration, 1973.

CHAPTER 4

CHEMICAL HEALTH HAZARDS

ANDERS ENGLUND, M.D.

The Construction Industry's Organization for Work Environment, Safety, and Health
Stockholm, Sweden

The overall chemical environment of the construction worker can be explained by analyzing the interactions between the chemicals of his external environment and those of his body. Chemicals in the external environment come from materials used; these may be breathed or swallowed, or they may contaminate the skin. Chemicals of the internal environment are the life substances by which bodily functions are carried on. Hazardous chemicals enter the body in three ways. (1) They can be inhaled in the form of dusts, fumes, gases, vapors, or mists; (2) they can be ingested through the mouth and digestive tract either directly or by contamination of food; or (3) they can be absorbed through the skin or affect the skin itself. Because these are the three routes by which hazardous substances enter the body, we first review how these bodily systems work.

4.1 THE SKIN

The skin acts as a barrier to the environment. The outer surface has a very rapid regenerating capacity. In general, the skin regenerates in response to wear; the more wear and tear received by surface skin, the faster it is replaced. In the tissue beneath the very outer surface there are pigments that act as a barrier against ultraviolet radiation from the sun. In the same tissue there are glands producing an oily substance called "sebum" which keeps the skin smooth and prevents drying of the surface. This protects from attacks by bacteria and fungi that are present everywhere. There are also sweat glands in the skin; these are important in producing the perspiration necessary for controlling body temperature. The dilation or constriction of the outer blood vessels of the skin also play an important role in temperature regulation. A number of sensory nerves are also present in the skin; these communicate essential information about temperature, pressure, and touch to the brain.

4.2   RESPIRATORY AND CARDIOVASCULAR SYSTEMS

The respiratory tract consists of the upper air passages of the nose and throat and the lungs. Air passages in the nose can remove particles from the inhaled air; these particles stick to the short hairs growing from the lining of the nose. Passage of air through the nose can be slowed by certain kinds of dust. In addition to its role in cleaning the inhaled air, the nasal passage also adjusts the temperature and moisture of the air before its arrival in the lungs.

The entire respiratory tract, including the nasal passages, is lined with mucous membrane. The cells on the inside walls of the tract have cilia, which are short, whiplike arms that are in constant motion. The cilia move together, causing a continuous wave of motion over the interior lining of the respiratory tract. Glands in the mucous membrane secrete mucus, a slippery substance that moistens and protects the membrane. The motion of the cilia causes particles caught in the mucus to move upward; these particles are transported out of the respiratory tract. This serves as a very efficient cleaning mechanism for the entire system (from the inner lining of the nose down to and including the lungs). The aerodynamic properties of the inhaled particles determine where they will land in the respiratory tract. Different types of substances collect at different points, some in the nose, some in the throat, and some in the lungs.

Studies have shown that inhaled fumes and gases can restrict or destroy the mobility of the cilia. Continuous exposure to substances such as tobacco smoke can profoundly alter the character of the mucous membrane, entirely destroying its function as a safeguard.

The larynx divides the food passage from the air passage in the throat: the esophagus goes to the stomach and the trachea goes to the lungs. The larynx also contains the vocal cords, the basic equipment for speech. Because of its location in the throat, the larynx can suffer

exposure to the harmful effects of both inhaled and ingested substances. Excessive intake of alcohol and tobacco are associated with
increased frequency of cancer of the larynx.

The trachea divides into two bronchi, one for each lung. As shown
in Figure 4.1, both bronchi branch continuously into smaller and
smaller passages; the smallest bronchi are called "bronchioli." The end
points of all these passages are called "alveoli," and it is here that
essential gases are exchanged (oxygen goes into the blood and carbon
dioxide comes out).

As indicated in Figure 4.1, particles larger than the diameter of the
smallest bronchioli will not be able to pass through them and penetrate

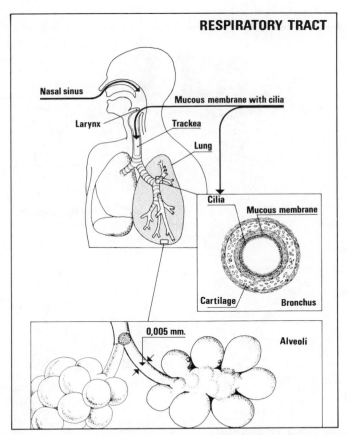

Figure 4.1 Antatomy of the respiratory tract with special emphasis on the wall of the
bronchus and alveoli.

the alveoli. It is important to note, however, that these larger particles can cause damage higher up in the respiratory system. A particle need not enter the alveoli to damage the lungs.

Gases are transported through the walls of the alveoli into extremely small blood vessels on the other side. As indicated in Figure 4.2, oxygen passes from the alveoli to the red blood corpuscles, where it forms a chemical bond with hemoglobin and is transported to every part of the body. The periphery of the body—muscles and other tissue—has a different pH from that of the lungs. This affects the hemoglobin's ability to bind the oxygen it carries, and the oxygen is released where it is needed. At the same time, the carbon dioxide that is produced as a waste material by metabolic processes is transported back to the

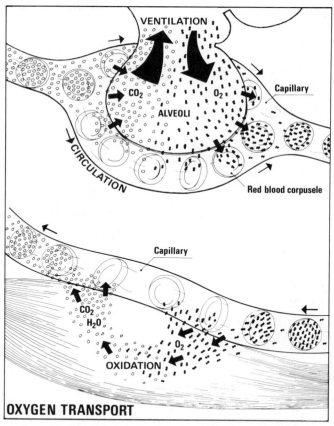

Figure 4.2 Transport mechanisms of oxygen and carbon dioxide.

lungs. Again, in the lungs the pH environment is such that the carbon dioxide is released, passes through the alveoli, and is expelled with the exhaled air.

There are several prerequisites for the transport of oxygen into the bloodstream through the walls of the alveoli. One condition, of course, is that the amount of oxygen in the inhaled air must be sufficient (in unventilated, confined areas, there may be too little oxygen). A second condition is that all parts of the lung are functioning. The lungs can be rendered partly or totally inactive by exposure to some chemicals. Lung function can be adversely affected in several ways. (1) Morphological processes can produce scars; this is the pattern of diseases such as silicosis and asbestosis. (2) Allergic reactions can cause obstruction of the smallest airways and impair the passage of oxygen to the alveoli. (3) Certain diseases affect the alveolus wall itself, impairing passage of oxygen to the blood as well as the excretion of carbon dioxide.

Another prerequisite for adequate oxygen transport involves the number of red blood corpuscles and their hemoglobin content. When the number of red blood corpuscles or the amount of hemoglobin in them is low, the transport capacity decreases. Carbon monoxide in the air also affects oxygen transport. Carbon monoxide binds to hemoglobin 200 to 300 times more easily than oxygen does, and leaves little hemoglobin free to carry oxygen. Such a condition leads rapidly to death because the cells of the brain require a constant, high level of oxygen for survival.

Toxic substances affect other vital components of the blood as well. The bloodstream carries leukocytes—white blood cells of various kinds—which protect against infection. One important group of leukocytes is the lymphocytes. These defend the body against invasion by two mechanisms: (1) by phagocytosis, actually surrounding or incorporating the invading organism or toxin; or (2) by antibody production. Toxic chemicals may affect leukocytes either by decreasing their production in the bone marrow or by increasing the rate at which they are destroyed in the bloodstream. Toxins may even change the distribution of white blood cells in the body. Severe bone marrow damage may result in a decrease in the numbers of all types of white blood cells. For example, exposure to ionizing radiation and benzene may damage bone marrow so severely that it fails to proliferate (a condition known as "aplastic anemia"). Leukemia, a malignant cancer that causes overproduction of leukocytes, might be another effect of such exposure.

The pump for the continuous circulation of blood is the heart. It has four chambers for distributing blood through the arteries and receiving returning blood through the veins. In addition to oxygen, fats and carbohydrates are essential to cells; these substances are delivered by the blood, along with essential trace minerals. Waste products of cell metabolism are transported away from the cells by the blood. During hard work, breathing and heart rate increase, in order to deliver more oxygen and energy to the working muscles. As a consequence, there is a higher concentration not only of oxygen but also of potentially hazardous substances in the lungs during work. Concentrations of any poisons that are present in the air will be higher in the blood during work because, as more air is taken in, more of every airborne substance is inhaled.

4.3 THE GASTROINTESTINAL TRACT AND THE URINARY SYSTEM

The gastrointestinal system is a continuous canal that starts with the mouth and ends at the rectum. Digestion of food takes place in the upper part of the system; absorption of the digested food and water take place in the lower part. Several organs that are not part of the digestive tract itself contribute to the digestive process. The pancreas and liver supply digestive juices to the small intestine through ducts. Both the pancreas and liver have other functions in the body: the pancreas produces insulin to control the concentration of available carbohydrates in the blood. Without sufficient insulin, diabetes develops. The liver has an important role in detoxification; it removes waste substances produced by normal metabolism as well as toxins taken into the body from the environment. These are carried to the liver in the blood, removed from the blood, and dumped into the small intestine along with the bile for elimination from the body.

As shown in Figure 4.3 and as previously mentioned, the larynx divides the entrance to the esophagus from the beginning of the trachea. The esophagus leads over to the ventriculus. These upper parts of the gastrointestinal system are very vulnerable to damage because they receive every ingested substance in unaltered form. Intake of acids or alcohol, for instance, can cause lesions in the esophagus wall. Large-scale studies show that differences in dietary habits—in drinking as well as eating—are associated with great differences in the occurrence of cancer in the upper parts of the digestive system. It has even been suggested that the involuntary ingestion of dust (from the swallowing of coughed-up inhaled particles) from the workplace can cause cancer of the ventriculus.

Food that is eaten mixes with digestive juices and passes further

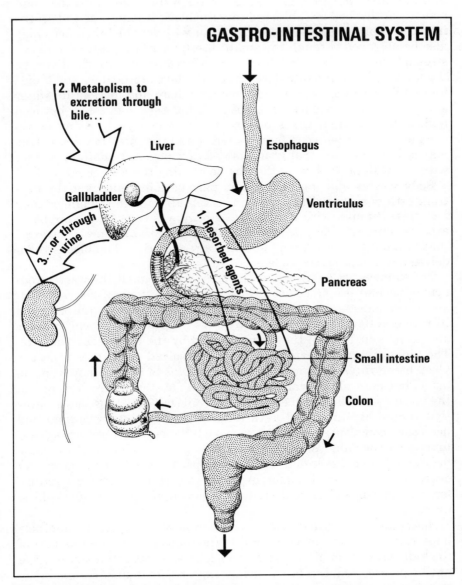

GASTRO-INTESTINAL SYSTEM

2. Metabolism to excretion through bile...

Liver

Esophagus

Gallbladder

3. ...or through urine

1. Resorbed agents

Ventriculus

Pancreas

Small intestine

Colon

Figure 4.3 The anatomy of the gastrointestinal tract with indication of uptake and excretion of agents and metabolites.

into the intestines. During passage through the small intestine, components of the food are absorbed into the blood through the wall of the intestine. Poisons swallowed with food or with coughed-up mucus may also be absorbed through the small intestine, passing into the bloodstream. The absorbed substances are all transported to the liver. In the lowest part of the digestive tract—the large intestine—the mixed bulk of food and digestive juices releases more and more of its water and, finally, is passed out of the body. It has been suggested that food high in fiber content will speed up this passage and, by doing so, reduce the contact between fecal material and the intestinal wall. This may reduce the time that irritants and carcinogens remain in contact with the wall of the bowel, thereby decreasing the risk of cancer.

Body wastes and ingested toxins may be eliminated via the gastrointestinal system. The liver can secrete these with the bile or they may pass through with food. Volatile body wastes or toxins might be excreted through the respiratory system. It is the kidneys, however, that handle the largest part of necessary excretion. The bloodstream delivers waste products to be excreted to the kidneys.

The kidneys act as filters for the body, filtering out the waste products of normal life processes (which can become toxic if not removed) and other substances in the blood. Blood flow through the kidney's filtration system is large, on the average of 1000 to 1200 cm^3/min, or about one-fourth of all the blood pumped by the heart every minute. The kidneys remove water and all the dissolved substances from the blood being filtered; some of these are wastes to be eliminated in the urine, but most of the water and dissolved substances are returned to the bloodstream. The kidneys selectively absorb only those substances that should be eliminated, returning to the blood every normal and necessary constituent that was removed. Some toxins can cause dysfunction of the kidneys, or physically damage them. Certain drugs have been shown to cause lesions and, after a period of decades, cancer may develop. Interactions between exposure to certain substances and infections have been reported to cause lesions in the filtering part of the kidneys.

Heavy metals used in the construction trade—lead, mercury, and, perhaps, cadmium—can cause lesions in the reabsorbing, selective part of the kidneys. Cancer of the urinary bladder, the organ that stores urine until it is excreted, can be caused by exposure to toxins. For example, the amines used in the rubber industry have been reported to cause bladder cancer. It is likely that contact between a carcinogen and the bladder wall, which may be prolonged, causes the cancer.

4.4 THE NERVOUS SYSTEM

Nerve cells are the most vulnerable to damage of any. All nerve cells are basically constructed the same way, whether they are part of the peripheral nervous system or part of the brain. In the periphery there are sensory nerve receptors; these receive signals from the environment and send signals to the brain (see Figure 4.4). From the motor area of the brain, nerve signals are sent through the spinal cord to the muscles. These signals result in coordinated action of many muscle cells, producing movements of the body (for instance, raising the arm and throwing a baseball). Damage to nerve cells can be caused either by lack of oxygen or by the direct effects of toxic substances.

Nerve tissue is rich in fat and it is usually the case that solvents (such as those in paint) accumulate in fat. The concentration of an inhaled solvent might become very high in nerve cells if solvent molecules accumulate in fat. If nerve cells are destroyed, the damage is irreversible because, unlike other body cells, nerve cells cannot be regenerated. Once they are destroyed or damaged they are gone for good.

The speed of nerve signals in the peripheral system has been shown to be reduced in people exposed to solvents. Changes in brain function have also been reported, although these are more difficult to assess. In some studies, electroencephalography (recording of brain electrical signals) has shown changed patterns owing to exposure to solvents. Other reports have shown decrements in psychomotor function, memory, and other brain functions. Studies have also indicated that exposure to solvents is related to the incidence of early retirement because of psychiatric disease. Some heavy metals, such as lead and mercury, have also been reported to cause lesions in both peripheral and central nerve tissue.

4.5 ADVERSE REACTIONS TO CONSTRUCTION MATERIALS

The work environment contains many chemicals, old and new, that can react unfavorably with human body chemistry. Many of the reactions are highly individual, because different workers' body chemistries react in unique ways to exposure to various substances. The effects of chemicals used in the construction industry fall into six categories: irritant, allergenic, dermatitic, toxic, fibrogenic, and carcinogenic. We review these one by one in this section.

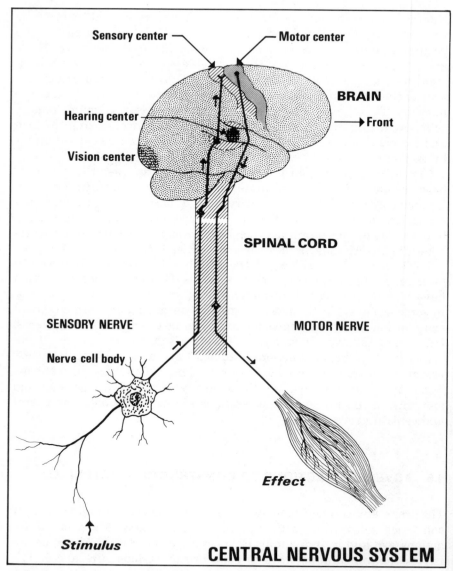

Figure 4.4 Basic structure of the nervous system with indication of sensory and motor nerve transport and some functional centers of the brain.

4.5.1 IRRITANT REACTIONS

Irritant substances affect the skin, the eye, or the mucous membranes of the respiratory tract. Strong acids can cause burns with blistering and subsequent scarring of skin, though these are not often a problem in construction. Weak acids cause redness, pain, and irritation, the effect depending on the acid, the duration of contact, and the availability of washing facilities.

More common irritants are alkalies, especially calcium hydroxide, one of the major components of cement. Burns on the knees caused by cement alkali are not unusual among cement workers who kneel on wet cement. Alkalies have the effect of dissolving the keratin layer of the skin. Without this protective layer, irritation progresses deep into the skin. This can cause secondary infection because it allows the bacteria always present on the skin surface to penetrate deep into the skin as well. Many solvents and resins are also primary skin irritants.

Alkalies such as ammonia, which is volatile, can irritate the delicate membranes of the eye. So, too, can many of the volatile solvents. The worst chemical effects on the eye result from splashes of alkali.

Irritation of the respiratory tract can arise from exposure to many dusts, fumes, and gases; the most serious problems are from ozone and oxides of nitrogen, which can produce delayed pulmonary edema. These gases are produced during welding, where the concentration can build up to high values, particularly in enclosed areas.

4.5.2 ALLERGIC REACTIONS

Allergic reactions are becoming more and more common among construction workers. The most important sites of these reactions are skin and lungs. Cement is the most common cause of skin sensitization among construction workers—the Cr^{6+} (chromium) ion can permeate skin tissue. Skin sensitization can also arise during handling of exotic woods such as teak, hickory, western red cedar, and so on. However, in second place behind cement, epoxy resins are a common cause of skin sensitization.

Once a worker is sensitized to a particular substance, he will react adversely to any exposure, however slight. The result is that he becomes unable to work with the substance that causes the allergy. Desensitization is very difficult, often impossible.

The most serious allergic reactions are those affecting the lungs; these can cause bronchial asthma. The substances most commonly associated with this reaction in the building industry are isocyanates.

Isocyanates have been used for some 30 years, primarily as an ingre-
dient in polyurethane paints. Current practice is to combine the isocyanate with the polyester base in order to bind it in a harmless form,
but there is always a danger of free isocyanate being present. The risk
is especially great in the spraying of polyurethane paint. Even though
the volatility of the isocyanate has been reduced, the possibility of
inhaling droplets that contain active isocyanate remains real and has
given rise to cases of asthma. As with skin sensitivity, once a worker
is sensitized he will react to very small quantities of the material. In
the case of TDI, for example, a worker may react to a concentration
of as little as 0.001 ppm (parts per million).

4.5.3 DERMATITIC REACTIONS

Skin disease is the most common occupational disease of construction
workers. It results from irritation or sensitization; among construction
workers, cement is the most common agent. Skin disease can also result
from the misuse of solvents and from some of the minor components
of paint (for example, some preservatives).

Solvents can dissolve the natural fat in the skin, leaving it open to
attack by bacteria or other external agents. There has been a great
deal of discussion about the use of barrier creams applied to the skin
before work. There may be some advantages to these creams, but they
may also give the worker a false sense of security (the feeling of wearing "invisible gloves"). Barrier creams make it easier to clean the skin
thoroughly at the end of the day, but their limitations should be recognized.

4.5.4 TOXIC REACTIONS

In the construction industry the hazard from toxic substances concerns
mainly toxic metals, toxic solvents, and certain gases. The most commonly encountered toxic metal is lead. A major class of toxic solvents
is the chlorinated hydrocarbons. Carbon monoxide and hydrogen sulfide are the major toxic gases of concern.

There is no problem in handling lead in its metallic form. Problems
arise when the metal is heated: the hotter the metal, the greater the
quantity of lead oxide fumes released. Until a few years ago, the most
common use of lead was as pigment in paint. White lead (lead carbonate), red lead (lead oxide), and lead chromate were, and in some
cases still are, used. There is small danger in applying lead-based
paints; the hazard is in contaminating the hands with lead, which is

involuntarily ingested with food. Much more hazardous is the removal of old lead-based paint by sanding or the fumes from burning it. In either case a fine dust or fumes are produced which are easily inhaled; the lead is then absorbed through the lungs into the blood, where it may cause anemia, colic, damage to the nervous system, and, finally, damage to the kidneys. Cadmium is another metal used in construction that can cause medical problems.

Solvents, especially the chlorinated hydrocarbon solvents such as trichloroethylene and methylene chloride, very often have a narcotic effect. Not only can fatal concentrations build up in confined spaces, but even smaller concentrations can cause a worker to lose his balance on a scaffold or roof beam.

The most toxic gases encountered in construction are carbon monoxide and hydrogen sulfide. Carbon monoxide is produced when carbon compounds burn with insufficient oxygen. The most common circumstances in which this happens in construction are when gasoline engines, braziers, or coal-fired boilers are used in confined areas. Toxic concentrations of carbon monoxide can easily build up in unventilated spaces; when this happens, carbon monoxide binds so much hemoglobin that the blood cannot deliver essential oxygen. Hydrogen sulfide gas may be present in any trench or pit; it is usually encountered in any hole dug where there has been a lot of plant material over the years. Hydrogen sulfide affects the breathing center of the brain and exposure can lead to rapid death.

A related problem that can occur in confined spaces is the simple absence of oxygen. The oxygen in the air may be used up in breathing or burning; it must be replaced by ventilation with fresh air. This is a good reason for testing the atmosphere prior to entering any confined area.

4.5.5 FIBROGENIC REACTIONS

"Pneumoconiosis" is the medical term for lung disease caused by habitual inhalation of irritant particles. Hazards from fibrogenic dusts are as old as the building trade. Stonemasons handling granite, sandstone, or any rock containing high proportions of silica are inevitably exposed to fine dust in the course of dressing the stone. Some modern methods of handling stone give rise to much larger quantities of dust, although they make the work easier.

Unless suitable wet methods are used to suppress dust, the worker is exposed to inhalation of fine particles. These, if they are sufficiently small (less than 0.005 mm), penetrate the alveoli of the lung. Once the

particles are in the alveoli, the body's defenses do not work; the result
is that fibrous tissue forms in the lungs. This ultimately leads to dust
disease—called "silicosis" if the inhaled dust is silicate. The important
factors are the nature of the rock (e.g., limestone is not silicate and
does not produce disease to the same extent), the size of the particles,
the concentration of particles in the air, and duration of the worker's
exposure.

Silica dust was a major hazard to construction workers in the past;
the most recent hazard has been exposure to asbestos. For the past 50
years asbestos cement has been used for covering outer walls of buildings. It has also been used in the form of inexpensive sheets that are
applied in interiors. Outside the construction industry, asbestos has
been used as an insulator for boilers, steam pipes, and, to a very considerable extent, in the shipbuilding industry. The classical disease
caused by inhalation of asbestos fibers is similar to the fibrous scarring
of the lung described for silicosis. In this case it is called "asbestosis."
Exposure to asbestos may also cause less specific reactions of the
pleural cavity: it may result in the production of excess fluid (pleuritis)
or an unusually high incidence of thickening and calcification of the
chest wall lining. Asbestos dust not only causes asbestosis, but also
has been incriminated as a cause of two different forms of lung cancer.

4.5.6 CARCINOGENIC REACTIONS

The hazards from carcinogens have been known for more than 200
years. Cancer of the scrotum in chimney sweeps was described in 1775
by Percival Pott. He recognized that the cause of the cancer was soot,
which contains carcinogens. Similar substances occur in oil, pitch, and
tar; exposure to these substances can cause cancer of the skin. Modern
lubricating oils are solvent-refined; it is unlikely that building workers
will be exposed to unrefined oils (although shuttering oils may be suspect). The use of asphalt ought not to give rise to cancer of the skin,
especially if natural asphalt is used. However, many so-called asphalts
are made from mixtures of pitch and tar, and these are likely to cause
trouble. As with other cancer-causing substances, there is a long period
between the time of exposure and the onset of the disease, as long as
40 years.

As previously mentioned, exposure to asbestos dust is clearly associated with the occurrence of at least two kinds of cancer of the lung
and chest. Asbestos can cause cancer of the trachea and bronchial tree,
and this particular risk is greatly enhanced if the worker smokes cig-
arettes. Asbestos dust can also cause cancer of the membrane that

covers the lung and lines the chest wall, the pleura. Cancer in the pleura is called a mesothelioma because the pleural tissue is derived from mesothelium. This particular tumor is not associated with smoking. A similar mesotheliomal membrane lines the abdominal wall and, even there, asbestos fibers have been found to cause mesothelioma. However, the mechanism probably involves ingestion of the fibers into the gastrointestinal tract rather than inhalation. Both these forms of cancer may arise 20 years or more after exposure to asbestos, even though there was no obvious damage to the lungs at the time of exposure.

It was believed until recently that mesothelioma was due to the dust of crocidolite (blue asbestos), but it now seems likely that the tumor can result from exposure to other forms of asbestos as well. Recent work shows that mesothelioma can arise in people who had no occupational exposure to asbestos but who were merely downwind from places where asbestos was used. It can also occur in people who live and work where the natural mineral fibers are found. This raises the whole question of whether man-made mineral fibers such as glass wool, rock wool, and slag wool might not have the same effect. The critical feature of a carcinogenic material is whether particles can be inhaled into the depths of the lungs. It is unlikely that fibers larger than 0.005 mm in diameter will get into the smallest lung spaces (see Figure 4.4). In general, man-made mineral fibers are of a diameter greater than 0.005 mm, though recent technology has produced smaller ones. Man-made mineral fibers do not split lengthwise into finer particles but, rather, tend to split horizontally. Therefore the risk of creating carcinogenic particles is less. More work must be done, however, to elucidate the role of man-made fibers in the human pathology of mesothelioma. There has so far not been any indication that cancer of the bronchial tree is associated with exposure to fibers other than asbestos. In addition to the different forms of cancer of the lung and of the abdominal wall, scientific reports and clinical findings suggest that cancer of the larynx and the digestive tract—especially of the stomach—is associated with occupational exposure to asbestos fibers.

Other occupational cancers among construction workers include the following:

1. Nasal sinus cancer may result from exposure to wood dust. This was first detected in the furniture industry, but has also been described in other woodworkers.
2. Angiosarcoma of the liver has been described for workers in manufacturing polyvinyl chloride (PVC). The suspicion arises that con-

struction workers handling articles made of PVC may be at risk. However, this risk is considered to be small, especially in light of recent
changes in the manufacture of PVC.

3. Research indicates that painters, exposed to solvents in their work,
may develop cancer of the esophagus and certain parts of the liver.

4.6 PRODUCT GROUPS AND ASSOCIATED RISKS

In analyzing chemical hazards, one approach is to group various products according to their composition and resulting health hazards. A
problem in any such grouping is that there are minor differences in
the compositions of related products. It is difficult to determine which
particular product in a group might be the safest one to use.

Many products that originally were used in controlled industrial
settings are now being marketed to people who are unaware of the
health risks involved. In construction work, the use of such products
saves time and energy, and produces satisfactory results—but at what
cost? Safe handling methods are generally neglected because often
workers are not aware that they are using hazardous substances. Many
of the products described below are commonly used by people who simply do not know that the materials they are handling can affect their
health for many years.

We analyze only the most frequently used types of materials. An
overview of the health hazards associated with the various materials
is given below in Table 4.1.

4.6.1   FORM OILS

Form oils are mainly composed of mineral oils, such as paraffins, with
various additives. Spraying of form oils produces a breathable mist
that might cause harm to the respiratory tract. Both spraying and
brushing involve contact with the skin and dermatitis is not an unusual result. Application by brushing is preferred to spraying from the
health viewpoint. Even with brushing the skin must be protected by
gloves. If application is to be made by spraying, calculations must be
done to assure that the maximum allowable concentration of oil mist
is not exceeded.

Exposure to cutting oils and oils used for lubricating machinery in
the textile industry has been reported to cause cancer of the scrotum
and, perhaps, lung cancer. So far we have no evidence of similar hazards from form oils. However, the time since form oil spraying was

TABLE 4.1 Threshold Limit Values (Exposure Limits) for some Pollutants Common in Construction Work

Pollutant	Limits for 8-hr Workday		Limits for 15 min Exposure	
	ppm	mg/m^3	ppm	mg/m^3
Asphalt fumes	—	5	—	10
Benzene[a]	10	30	—	—
Butane[b]	600	1430	750	1780
Calcium carbonate/marble	—	10[c]	—	20
Carbon monoxide	50	55	400	440
Carbon tetrachloride (skin)	10	65	20	130
Chlorine	1	3	3	9
Coal tar pitch fumes[a]	—	0.2	—	—
Ethyl alcohol	1000	1900	—	—
Graphite (synthetic)	—	10[c]	—	—
Limestone	—	10[c]	—	20
Liquified petroleum (LPG)	1000	1800	1250	2250
Methyl alcohol (skin)	200	260	250	310
Naphthalene	10	50	15	75
Nitrogen dioxide	5	9	—	—
Oil mist, mineral	—	5	—	10
Ozone	0.1	0.2	0.3	0.6
Paraffin wax fumes	—	2	—	6
Particulate hydrocarbons	—	0.2	—	0.2
Plaster of Paris	—	10[c]	—	20
Propane[b]	—	—	—	—
Propyl alcohol	200	500	250	625
Sulfur dioxide	2	5	5	15
Toluene (skin)	100	375	150	560
Trinitrotoluene (TNT)	—	0.5	—	—
Turpentine	100	560	150	840
Welding fumes[d]	—	5	—	—
Wood dust	—	5	—	10
Xylene (skin)	100	435	150	655

[a] Carcinogens
[b] Asphyxiants; that is, oxygen must be available at at least 18% to support life.
[c] When quartz < 1%.
[d] Ozone, carbon monoxide, iron, manganese, silicon, chromium, nickel, flourides.

introduced might still be too short for the detection of cancer risk. Risks from spraying form oil are increased if lead is added to the oil as an anti-rust agent.

4.6.2 CONCRETE AND ITS ADMIXTURES

As described previously, the alkalinity of concrete (caused by sodium hydroxide, sodium silicate, calcium silicate, etc.) causes lesions on the skin and, if splashed, will damage the eye. Real burns may sometimes be seen, especially on the knees. The major problem in cement handling, however, is the condition called "cement eczema," which is an allergic reaction to Cr^{6+}. The Cr^{3+} ion is less water soluble and less biologically active, so reduction of the chromium to the Cr^{3+} form reduces the hazard. Such a chemical reaction is readily effected by adding ferrosulfate to the cement. In Sweden, ferrosulfate is used for this purpose at all construction sites.

Conventional admixtures for accelerating are calcium, magnesium, or aluminum chloride. Hydrophobic agents are generally soaps. Neither these components nor fillers are very likely to cause major health problems. Very little is known as yet about other chemicals added to the cement to change its performance, for example, to make it more fluid or to make it harden faster. Although the chemical dangers are unknown, it is clear that the more fluid cement can be pumped and unloaded with far less difficulty. And, since less vibration is required, the risk of health problems caused by vibration is reduced.

4.6.3 INSULATION MATERIAL

Materials for fire and thermal insulation are generally fibrous. Originally, various kinds of asbestos fibers were used. Since the grave health hazards associated with asbestos were recognized, there has been a change to man-made mineral fibers such as mineral wool, glass fiber, and, for some applications, cellulose fiber. The problems of fibrogenicity and carcinogenicity of different fibers have already been dealt with in this chapter. Although the health record of the man-made fibers is not yet entirely clear, there is reason to assume that it is substantially better than asbestos.

Different application methods lead to substantially different exposure to fibers. The highest fiber counts result from spraying; installation of cement-bound fiber sheets releases only a few fibers. Aside from respiratory tract damage from fiber inhalation, a prominent effect of mineral wool and glass fibers is itching.

Thermal insulation may be done with polyurethane products. As with polyurethane paint, spray application increases the risk of damaging concentrations of free isocyanates in the atmosphere. This can cause respiratory tract and skin damage, manifested by varying degrees of irritation or even allergic reactions. Heating causes the polyurethane to deteriorate, releasing noxious gases. In a confined area a plumber or pipe fitter might be heavily exposed during welding on pipes insulated with polyurethane.

4.6.4 ASPHALT

Applied cold, in sunlight, asphalt can cause skin irritation. Applied hot, as it frequently is, asphalt can cause more severe skin irritation and burns. Heated asphalt releases fumes that can cause severe irritation of the respiratory tract. For other occupational groups and exposure conditions, the polycyclic aromatic hydrocarbons released from hot asphalt have been found to cause lung cancer. Whether this is the case among roofers is not clear, but it is not unlikely. It has been reported that chronic bronchitis is found more frequently than normal among roofers.

4.6.5 CAULKING AGENTS AND SEALANTS

Caulking agents and sealants may be liquids applied by spraying or soft solids applied with tools. The more solid but less elastic types of material are mainly composed of polyacrylate; this often contains unreacted monomers. Such monomers are irritating to the eyes and skin, and some central effects are possible. However, the fibers and solvents in these products are not likely to have harmful effects with normal use. The more elastic types of caulking and sealants are generally composed of rubber and solvents such as methyl ethyl ketone and methyl isobutyl ketone. Both of these are irritants to the skin and upper respiratory tract.

The more liquid caulking and sealant products are composed of isocyanates that have been chemically reacted with other compounds. Fluorocarbons are often used as propulsion agents in sprays. Isocyanates are likely to be inhaled, particularly when these products are used in hot, dry places. The dust particles generated during grinding and smoothing of excess caulking material contain isocyanates. The isocyanates are allergenic and might also cause eczema or asthmatic attacks. Heating of these products (for instance, in cleaning tools) will lead to release of noxious gases. Noxious gases result from the disintegration of the fluorocarbon propulsion agent; even cigarette smoking during spraying will produce enough heat to deteriorate the fluorocarbon. Heating of sprayed materials produces the same effect.

4.6.6 ADHESIVES

Adhesives have varied compositions depending on the purpose for which they are used. A modern adhesive generally consists of a polymer, such as acrylate, dispersed in a solvent or water. Added to this are preservatives and, generally, some filler. Some adhesives are based on rubber compounds such as neoprene, and in these adhesives, the solvent content can be quite high. Most of the health hazards associated with these products are due to irritant effects on the eye and upper respiratory tract. The irritation is produced by unreacted monomers, such as acrylic monomer, and by the effects of solvents, such as toluene and xylene. In addition to these effects, exposure to such solvents may pose a risk to the central nervous system. Sensitive people may also react to the preservatives.

Another group of adhesives have a two-component structure with a polyurethane resin and isocyanate (MDI). The main hazard is the allergenic property of the isocyanates; especially high levels are released when these adhesives are heated (as when tools are cleaned). Skin reactions and respiratory discomfort can result from use of these products; asthmatic attacks have been known to occur as well. Another two-component type of adhesive is based on an epoxy resin and a polyamine. The epoxy resin can penetrate the skin; the lower the molecular weight of the epoxy, the more easily it penetrates. The polyamine has a high pH and can irritate the skin. To avoid skin problems the two components should be mixed in disposable units and the hands should be cleaned frequently during handling. Inhalation of dust produced by grinding the hardened epoxy resin should be avoided.

4.6.7 PLASTERS AND DRYWALL

Modern, fast-curing plasters are of a two-component polyester type. Unsaturated polyester reacts with styrene and peroxide to start the hardening process. Styrene evaporates during hardening and, when the product is applied on large surfaces such as floors (to be made smooth for tile), substantial amounts of styrene can be inhaled. Styrene is taken into the body through the skin and through inhalation. It has a pronounced irritant effect; effects on the central nervous system have also been reported. There is some reason to suspect that it might produce cancer.

Drywall products are composed of calcium carbonate and talc, along with preservatives and antibacterial agents. Nowadays, fillers such as talc are free from asbestos contamination, but this was not always the case. The unexpectedly high frequency of asbestos-related lung disorders among painters might be due in part to application and sanding of drywall.

4.7 DISEASES AMONG CONSTRUCTION WORKERS

How can we determine the real health risks involved in construction work? There are two ways to study occupation risks from exposure to chemical substances. One is to perform toxicological studies in the laboratory; the other is to look at health statistics. Laboratory studies can tell us what to expect from a product under certain conditions of use. Health records and statistics can show what risks were present and what the actual results of exposure were under real-life conditions. It is clear that all possible measures must be taken to perform adequate laboratory tests before products are put to general use; otherwise, the worker becomes a guinea pig. It is equally clear that, in addition to the laboratory assessment of toxicity, we must keep track of what happens during repeated exposures in actual work situations. This should be done through careful monitoring of health records of workers.

4.7.1 EPIDEMIOLOGICAL PRINCIPLES

Few studies have been performed and published on health and disease among typical consumer groups like construction workers. On the other hand, the possible health consequences of handling certain materials have been studied among workers involved in the manufacture of these materials. Inferences can be drawn from the studies of manufacturing workers and applied to the people who use the products. Actual risk to construction workers (and others who use the products) may be higher than to workers who manufacture hazardous materials: exposure may be greater or the material in question may be used in combination with other substances that make it more dangerous still. Many chemicals are only moderately toxic in themselves, but become dramatically more toxic in the presence of other substances.

Studies of construction workers' health as a function of exposure to chemicals are extremely difficult to do. One problem is that no one knows of or keeps records of exposures. A few experts do know the composition of the materials used and, in a few places, there are programs measuring exposure levels. But working conditions and tasks change often in the construction industry, so it is difficult to assess any worker's total exposure.

Another problem in studying worker health is that construction workers are an extremely unstable work force. Employees change from one building project to another, from one employer to another, and from one part of a country to another, depending on where work is available. In several European countries they even move from one country to another. In France and West Germany, for example, more than half of the construction work force might be from another country— these are migrant workers who stay for a comparatively short time. It is impossible to follow such workers through a full working life to see what diseases they develop. To complicate the picture, diseases may result from work in other occupations, for many workers do other kinds of work besides construction. Those who develop symptoms or diseases while working are likely to leave construction as soon as they can; information about them is simply not available.

All these conditions act as selection mechanisms and make studies difficult to do and interpretation even more difficult. In principle, the effects and trends that can be established should be regarded as underestimates of the real case; only the tip of the iceberg is visible because of the severe selection that takes place. In practical terms, the fact that nobody talks about health hazards in the construction workplace should not be taken to mean that there are none. It might simply result from the fact that nobody looked into the matter. Or that those who looked into the matter did not find anything definite because what should have been found had disappeared: workers who became sick left the work force or moved away from the area of exposure back to their places of origin.

4.7.2   RESULTS OF EPIDEMIOLOGICAL STUDIES

Two general surveys of causes of death in different occupational categories are described. In England and Wales, since the last century, the Register General has issued a report every tenth year on deaths in each occupational group. The most recent one covers the period 1970–1972. Death due to accidents plays a prominent role for both construction workers and painters. Malignant neoplasms of the trachea, bronchi, and lungs are more frequent than normal in both these occupations. Construction workers also have unexpectedly high frequencies of cancer of the stomach and lips.

In the United States, Samuel Milham has presented a comprehensive but less analytically elaborate study. It shows the proportional mortality in a large number of occupational groups for 1950–1971. Again, incidence of accidents and of cancer of the lung and of the stomach are found to be higher than normal among bricklayers, stonemasons, tilers, and building and construction contractors. Cancer of the urinary bladder is more frequent among construction workers than other occupational groups. For painters, the study reports an abnormally high incidence of lung cancer. Roofers and slaters have increased mortality from lung cancer, asthma, and pulmonary emphysema. Plumbers, pipe fitters, and asbestos and insulation workers have a general high incidence of malignant diseases, mainly cancers of the respiratory tract. A number of nonmalignant diseases are also found to be proportionally more frequent in construction workers: diabetes, gastric and duodenal ulcer, cirrhosis of the liver with alcoholism, gallstones, and acute pancreatitis. It should be noted that these are based on causes of death and not on the mere occurrence of diseases.

Studies on insulation workers from union records in New York and in other American and Canadian areas have been reported by Irving Selikoff at the Mount Sinai Hospital in New York. These workers experienced higher than normal mortality from cancers of the respiratory tract (lung cancer and mesothelioma of the pleura) and from cancers of the gastrointestinal tract (mainly stomach cancer and mesothelioma of the peritoneum). These studies have shown the great importance of extended follow-up of workers exposed to hazardous materials. The increased cancer-related death rate does not appear until 20 or 30 years after the first exposure. These studies also confirmed findings from other asbestos-exposed groups showing excess deaths due to cancer of the larynx and to asbestosis.

Ten years ago in Sweden, an occupational health service, BYGGHÄLSAN, was set up to serve the construction industry and its workers all over the country. In addition to ordinary, day-to-day preventive health and safety work and the treatment of patients, the health service has done some specific epidemiological studies on causes of death and cancer incidence in selected occupational categories. Both death and disease statistics show that painters have higher than normal incidence of cancer of the esophagus and certain parts of the liver and certain leukemias. This pattern is shared by some other occupational groups exposed to solvents. These workers also experience higher rates of respiratory tract cancer. Mortality statistics also show excess mortality due to stomach ulcers and due to asthma and emphysema. Deaths related to alcoholism without cirrhosis of the liver are found in this group, although the combination seems somewhat unlikely.

Mortality studies and studies of cancer show that plumbers and pipe fitters have abnormally high rates of respiratory tract cancer. Lung cancer is most common, but mesothelioma of the pleura and cancer of the larynx and nasal sinuses also occur.

Some general construction work categories studied by the Swedish health service so far have shown only slight deviations from the expected incidence of respiratory and gastrointestinal cancers. Bricklayers have a definite higher-than-expected incidence of lip cancer. No mortality figures are yet available for these general construction work categories.

4.7.3 EXPOSURES

Table 4.1 presents threshold limit values (or exposure limits) for some chemical substances common in construction work (General Industry Safety and Health Regulations, Part 1910, *Federal Register*, June 27, 1974). Depending on the amount of exposure during a day, there are different threshold limit values (TLV). In the table, threshold limit values are given for 8-hr and 15-min exposures.

For the case that a worker is exposed to a mixture of contaminants, the threshold limit values must be added. Assume exposure to liquefied petroleum gas at 500 ppm, methyl alcohol at 45 ppm, and propyl alcohol at 40 ppm. The three TLV's are 1000, 200, and 200 ppm (see Table 4.1). Then the time weighted average (TWA) can be calculated:

$$\text{TWA}_{\text{mixture}} = \frac{C_1}{\text{TLV}_1} + \frac{C_2}{\text{TLV}_2} + \frac{C_3}{\text{TLV}_3}$$

where $\text{TWA}_{\text{mixture}}$ = equivalent TWA mixture exposure (maximum

of 1 permitted)

C = concentration in ppm

For the numbers above $\text{TWA}_{\text{mixture}}$ is calculated to be 0.925. Since 0.925 is less than 1, the exposure is legal.

4.8 SUMMARY

The health of the construction worker depends on his internal and external chemical environment. Individual reactions to hazardous substances used at work vary because every worker's internal system is different (and the dangers are affected by previous exposures and exposure to other chemicals at the same time). Some of the various materials encountered in construction work can damage or destroy essential organs and life processes over time. Hazardous chemicals may be inhaled, ingested, or absorbed through the skin.

The skin is the first barrier to toxins. Skin diseases are very common among construction workers who handle cement, solvents, abrasives, acids, and alkalies.

The respiratory tract is the target of all inhaled toxins and particles. It consists of all air passages from the nose to the tiniest alveoli in the lungs. The alveoli are the sites of oxygen and carbon dioxide exchange; damage to these structures results in an inability to take in essential oxygen. The respiratory tract has defenses against inhaled particles, but these defenses may be destroyed by exposure to various particles, gases, and dusts over time. Exposure to fibrogens at work can cause scarring of lung tissue that prevents lung function. Allergic reactions obstruct small airways and impair the passage of oxygen. Oxygen transport is also impaired by the presence of carbon monoxide in the air.

Toxic substances can tie up hemoglobin, blocking oxygen transport. Some substances affect white blood cells as well, by destroying them in the blood or by damaging bone marrow (where they are produced). Certain leukemias can result from exposure to environmental toxins.

Hazardous substances may be taken into the body with food or with the swallowing of coughed-up material. The gastrointestinal system might absorb the substance into the blood or eliminate it with normal body wastes. The kidneys act as filters for the blood, so all the substances carried in the blood end up concentrated in the kidneys. Wastes are eliminated after storage in the urinary bladder, so the bladder may be exposed to concentrated toxins in the urine.

Nerve cells are extremely vulnerable to damage from toxins and, generally, cannot be replaced by the body if they are destroyed. Nerve tissue is rich in fat and, unfortunately, many toxins are stored and concentrated in body fat. Solvents have been shown to be related to nerve damage, decrements in brain function, and early retirement due to psychiatric disease. Some heavy metals have also been reported to damage nerve tissue.

Table 4.2 summarizes the product groups that may be hazardous to worker health. Methods for safer application and handling are available or can be developed. But it is essential for workers and supervisors to realize that there are hazards involved in the use of these common substances, especially in long-term, day-to-day exposure to them. Once

TABLE 4.2 Construction Materials Causing Occupational Diseases

Construction material	Disease				
	Skin	Respiratory tract	Lungs	Eyes	Nervous system
Form oils	Irritation				
Cement (chromium)	Lesions, eczema, burns	Irritation			
Insulation					
Mineral wool	Irritation				
Glass wool	Irritation				
Isocyanate	Damage	Damage	Asthma		
Asphalt	Irritation, burns		Bronchitis, cancer?		
Caulking and Sealants					
Polyacrylate	Irritation			Irritation	
Methyl ethyl ketone	Irritation	Irritation			
Methyl isobutyl ketone	Irritation	Irritation			
Isocyanate	Eczema		Asthma		
Adhesives					
Acrylic monomers	Irritation	Irritation		Irritation	
Toluene	Irritation	Irritation			Damage
Xylene	Irritation	Irritation			Damage
Isocyanate		Discomfort			
Epoxy resin	Irritation				
Polyamine	Irritation				
Plasters, drywall, putties (Styrene)	Irritation	Irritation			

this is taken into account, means can be devised to lessen the risks involved.

It is difficult to assess the effects of exposure to chemicals on the health of construction workers. The studies that have been done show that construction workers generally have higher than normal incidence of death from cancers of the respiratory and gastrointestinal systems.

CHAPTER 5

WORK PHYSIOLOGY

JAMES C. MILLER, Ph.D.

Human Factors Research, Inc.
Goleta, California

STEVEN M. HORVATH, Ph.D.

Institute of Environmental Stress
University of California, Santa Barbara

This chapter is an introduction to the functioning of the human body as it performs physical labor. More extensive treatments of work physiology are found in texts by Åstrand and Rodahl (1970) and Lange–Anderson et al. (1978).

In simple terms, the body is an engine that provides muscle power to perform work. Like any engine, it must operate within the limits set by the laws of physics. The body uses oxygen, which enters through the lungs, to burn fuel, which has entered through the digestive system. The energy is converted into the kinetic and chemical energies required for sustaining body functions. The conversion is complicated, but it has been studied in depth and is well documented. It is known as metabolism and occurs within the cells of the body. Before metabolism is discussed, the jargon of work physiology will be presented in terms of a standard system of units.

5.1 UNITS

The standard system is the International System of Units (SI units). Conversions of commonly used units to SI units are presented in Table 5.1.

TABLE 5.1 Conversion of Common Units to SI Units of Measure

Quantity	Common units	Factor	SI units
Time	second, minute, hour (sec, min, hr)	× 1	s, min, h
Age	year (yr)	× 1	annum (a)
Distance (height)	foot (ft)	× 0.30480	meter (m)
	Mile (mi)	× 1.6093	kilometer (km)
Velocity (speed)	ft/sec	× 0.30480	m/s
	mi/hr, mph	× 1.6093	km/h
Volume	quart (fluid, U.S.) (qt)	× 0.94633	liter (l)
	gallon (U.S.) (gal)	× 3.7854 × 10^{-3}	m^3
Force (weight)	pound (lb)	× 4.4482	newton (N)
	kilogram (kg)	× 9.8066	N
Energy (work)	ft-lb	× 1.3558	joule (J)
	kg-m (kgm) or kilopond-m (kpm)	× 9.8066	J
	kilocalorie (kcal)	× 4186	J
Power (work rate)	horsepower (hp)	× 745.70	watt (W)
	kgm/min, kpm/min	× 0.16344	W
	kcal/min	× 69.767	W
Temperature	°Farenheit (°F)	5/9(°F-32)	°Celcius (°C)
	°F	5/9(°F-32) + 273.16	°Kelvin (°K)
Pressure (at 0°C)	Millimeters of mercury (mm Hg or torr)	× 133.32	Pascal (Pa)
	atmospheres (atm)	× 1.0132 × 10^5	Pa
	lb/in.^{2}	× 6.8974 × 10^3	Pa

Source: Hodgman, C. D. (ed.). *Handbook of Chemistry and Physics*, 44th ed. Cleveland: Chemical Rubber Publishing Co., 1962.

Volume (V) the work physiologist measures volume in liters (l) and milliliters (ml; 1 ml is 0.001 liter). Flow, or volume per unit of time, is indicated by a dot above the abbreviation. Thus oxygen (O_2) uptake by the body is indicated by the symbol $\dot{V}_{O_2}$. Oxygen uptake is usually measured in l/min or, corrected for body weight, as ml/min per kilogram of body weight.

Force (weight) is normally measured in kilograms (kg). *Energy expenditure* (work) is measured in joules (J), the energy expended when a 1 kg mass is moved 1 m by the force of 1 newton. However, the work physiologist has traditionally measured work in a related unit, the kilocalorie (kcal), which represents the energy expended when 1 liter of water is heated 1°C (from 15 to 16°C). The kilocalorie is the "Calorie" of the weight-conscious dieter. Determinations of the caloric equivalents of foods and, concomitantly, the amount of oxygen required to burn them, are performed in a device called a "calorimeter" and by a process known as "direct calorimetry." The estimation of human energy expenditure through the measurement of the amount of oxygen consumed in order to burn fuel is known as "indirect calorimetry." When the body burns 1 liter of oxygen, it generates 4.83 kcal of energy. *Power* (work rate) is measured in watts (W) in the SI, a watt being 1 J/s. The work physiologist has, in the past, used a related unit, the kilogram-meter/minute (kgm/min) or its equivalent, the kilopond-meter/minute (kpm/min) to measure work rate.

Temperature is normally measured in degrees Celsius (°C). *Pressure* is measured as the Pascal (Pa). However, blood pressure is traditionally measured in terms of the height of a column of mercury, as millimeters of mercury (mm Hg) or torr.

Heart rate is measured in beats/min.

5.2 METABOLISM

There are two processes in energy metabolism that must be considered: energy conversion inside the body and the muscular work performed (see Figure 5.1). In Figure 5.1, fuel is shown entering the body (the vertical rectangle) in the form of food energy. The utilization of this energy for work and for producing heat is shown in the figure. At the top of the rectangle, owing to the thermodynamic energy exchange, 5% of the energy is lost as heat.

The next step shown is the conversion of food energy into the high-energy compound adenosine triphosphate (ATP). ATP serves as a fuel transport mechanism. It uses the energy available in food to synthesize

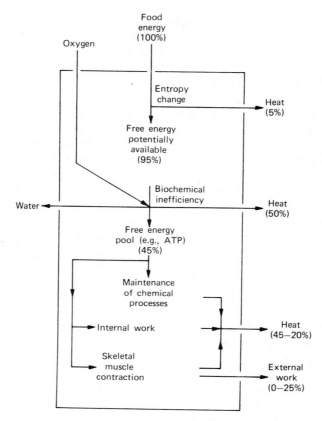

Figure 5.1 The distribution of food energy within the body.

high-energy chemical bonds. Later the ATP can release that chemical energy to fuel such processes as muscle contraction. The ATP-synthesizing process is about 50% efficient: about half of the total food energy is lost as heat.

The ATP energy is used in three different processes. The first, "maintenance of chemical processes," represents the synthesis and maintenance of high-energy bonds in chemical compounds other than ATP. Second, internal work such as neural processes and muscular contractions that maintain blood flow and breathing use another portion of the available energy. Finally, some of the ATP-stored energy is used for muscular work.

About 25% of the potential energy that enters the body in the form of food can be used for muscular work. This is the upper limit of the energy efficiency of the human body. The remaining energy is converted to heat. The 25% maximal gross efficiency of the body as a machine exceeds the efficiency of a steam engine, about equals the efficiency of an internal combustion engine, and is inferior to the efficiency of an electric motor (Brown and Brengelmann, 1965).

The amount of muscular work performed can be assessed by measuring the amount of oxygen used. Information acquired in this way allows the comparison of physical difficulty among many different kinds of tasks and the establishment of acceptable work loads for different occupations (Singleton, 1972). A common way to measure oxygen use is to analyze the oxygen content of the exhaled air and compare it with the inhaled air. This analysis can be performed with special instruments. Once the oxygen uptake value is obtained, it can be converted into kilocalories to assess energy expenditure (metabolic rate).

5.2.1 BASAL METABOLISM

In order to sustain life, there must be a minimum energy conversion rate. This is known as the basal metabolic rate (BMR). BMR can be calculated by measuring the oxygen consumption, $\dot{V}_{O_2}$. The following formulas can be used to estimate BMR for males and females (International Commission on Radiological Protection, 1975):

$$BMR_{(males)} = 66.5 + 13.8(\text{weight}) + 5.01(\text{height}) - 6.8(\text{age})$$
$$BMR_{(females)} = 66.5 + 9.6(\text{weight}) + 1.8(\text{height}) - 4.7(\text{age})$$

The energy is calculated in kilocalories per 24 hr, weight is in kilograms, height is in centimeters, and age is in years. Figure 5.2 shows metabolic rates calculated for males and females using the above equations.

As can be seen in Figure 5.2, BMR decreases with age and is substantially lower for females than for males. The basal metabolic rates calculated in the equations represent estimates for 24 hr. However, the rate varies during the 24-hr period because there are several natural basal conditions. The lowest values of BMR are reached in the hours immediately preceding dawn (Miller and Helander, 1979; Miller and Horvath, 1976, 1977). This is due to the 24-hr rhythmic properties of the brain, which regulate the metabolism so that various body functions are most active during the day and least active at night. Accordingly, values of BMR measured during the day are always higher, regardless of whether the person is asleep or not.

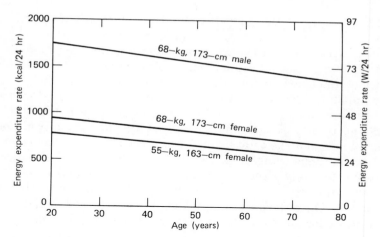

Figure 5.2 Basal metabolic rates for standard size male and female of equal size and for female of more typical size, as a functions of age.

5.2.2   METABOLISM DURING WORK

Some confusion exists about the meaning of the term "work" when it is applied to muscle contraction. If an individual holds a 10-kg weight horizontally at arm's length, an engineer might claim that no physical work is being done, since the weight is stationary. In fact, the joints of the arm are loaded by a moment arm (force x length) and compensating work must be performed by the muscles of the arm.

A second point worth noting is that, although a muscle can convert chemical energy into mechanical energy, the reverse cannot occur. For example, consider lifting a weight vertically from a table and then replacing it. As the weight is lifted, muscles do physical work on the weight, increasing its potential energy. As the weight is lowered, it does an equal amount of physical work on the muscles, but this work cannot be converted into chemical potential; it is lost as heat. The body is simply unable to use any energy source other than food energy (Brown and Brengelmann, 1965).

The mere anticipation of physical work results in an elevated energy conversion rate prior to the performance of the work (Cannon's "fight or flight" syndrome). Once work has begun it takes some time for the metabolism to "catch up" with the energy expenditure of the muscles engaged in work. In fact, metabolism does not reach a stable, adequate level until several minutes after work has begun. The amount of time taken to reach this stable, adequate level depends upon how hard the work is. Figure 5.3 shows the time delay for the metabolic processes. The metabolic rate, or oxygen uptake, does not increase suddenly at the onset of work. Rather, there is a gradual, smooth increase in oxygen uptake.

During the initial portion of work, the muscles use a type of energy that does not require oxygen. This type of energy production is known as "anaerobic" (without oxygen) metabolism. It is possible to perform a short task, such as a 100-m sprint, using this kind of energy production primarily.

Slowly, as the oxygen uptake increases, the body can use aerobic or oxygen-requiring fuel (ATP). Anaerobic metabolism is limited in its efficiency: it uses nearly 20 times more food–fuel than does the aerobic process. It produces a waste product, lactic acid, which may accumulate in the working muscles rather than being carried away by the blood. Eventually, lack of available energy stores, lack of fuel, and accumulation of lactic acid in involved muscles lead to the voluntary or involuntary cessation of work. It is generally believed that the accumulation of lactic acid results in temporary muscle ache.

Returning to Figure 5.3, it can be seen that the metabolic rate eventually stabilizes during work. This steady-state level represents the body's aerobic response to the demands of the increased work load. When the work ceases, the oxygen uptake returns slowly to the resting level prior to work. The slow return after work represents the oxygen debt incurred during the onset of work (area A in Figure 5.3) being repaid (area B).

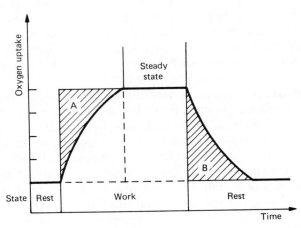

Figure 5.3 Oxygen uptake at the onset of, during, and after work. Composition of metabolic delays: A = oxygen debt; B = repayment of oxygen debt during rest; A ≅ B.

5.2.3 ENERGY EFFICIENCY

With a general understanding of internal energy conversion processes and of the nature of external work, an example of the calculation of human work efficiency can be discussed. A man of average height (173 cm) and average weight (68 kg) is doing plaster lathing work. This task imposes about 20 kcal/hr (23 W) of external work. His resting metabolic rate just prior to working is about 80 kcal/hr (93 W). The steady-state energy expenditure for this task is 180 kcal/hr (209 W; Altman and Dittmer, 1968). This value can be calculated by measuring his oxygen consumption during work. The worker's *net* efficiency at putting up lath is calculated as the ratio of external work to energy expenditure *increase*, or 20/100 = 20%. The efficiency for this kind of work is therefore less than the maximal 25% mentioned earlier.

5.3 INDIVIDUAL DIFFERENCES IN WORK PERFORMANCE

The capacity for physical labor varies greatly from person to person. Some of the major reasons for this are the level of physical fitness, nutritional status, age, sex, and health. When two equally skilled individuals perform identical physical tasks, the individual with the lower work capacity (a lower level of aerobic physical fitness) will find the task more difficult. The less fit individual is pushed closer to his work limit by the task and thus tires more quickly. In order to establish limits for physical workload for a particular task, the maximal physical workloads for a number of different individuals must be compared.

The best, though not fully adequate, way to measure work capacity is the maximum capacity to use oxygen, max $\dot{V}_{O_2}$. There is no adequate way to measure anaerobic work capacity. However, anaerobic work is seldom performed during all-day work.

Maximum oxygen uptake is determined for many different reasons: to diagnose cardiac problems, to evaluate the potential of athletes, or to determine the capacity to perform a particular job. In order to measure a normal individual's functional limits, the $\dot{V}_{O_2}$ must be measured during maximum work. To test an athlete's potential, such a test must be performed for an extended period of time, so that test duration resembles the real task. For example, a marathon runner works at 75 to 80% of max $\dot{V}_{O_2}$ for over 2 hr (Maron et al., 1976).

Several different tests are currently used in exercise or stress testing. They employ stepping, bicycling, running, or walking as the physical task. A common submaximal test used by physicians is the Master Two

Step Test with concurrent measurement of the electrocardiogram. In the Harvard Step Test, the subject steps on and off a 16-in.-high (41 cm) platform 30 times/min for 5 min, and the recovery heart rate is measured. This test, which may lead to maximal effort, is useful in the diagnosis and monitoring of heart disease and for monitoring the effect of physical training programs.

Other tests are designed to measure maximal aerobic capacity. Maximal tests are usually preceded by a period of submaximal testing in order not to suddenly overload the various physiological systems. Heart rate, blood pressure, lung ventilation, oxygen uptake, and lactic acid levels are usually measured during and after exercise tests. Sometimes a recovery period of 5 to 15 min is required in order to measure oxygen debt (see Figure 5.3).

If the individual's maximum aerobic capacity has been determined, it is possible to predict his potential to perform physical work. The higher the individual's maximum aerobic capacity, the larger the amount of available energy. However, there is no guarantee that such tests can be applied to a specific industrial job. Technique (skill), strength, and motivation can be equally important to job performance. Nonetheless, a test of the limits of physical performance capacity is often useful to ensure that the work capacity is not exceeded by job requirements.

5.3.1 EFFECTS OF AGE AND SEX ON WORK CAPACITY

The aerobic work capacity peaks in the late teen years. An average male has a maximum oxygen uptake of about 3.5 l/min at this age. This is the equivalent of 1179 W of energy. This figure can be translated to maximum workload using the 20% net work efficiency figure and the basal metabolic rate equations given earlier. For a 173-cm tall, 68-kg man, maximum work capacity is 236 W (of external work). The 23-W lathing task thus represents a 9.7% (23/236) relative work load for this individual.

Work capacity decreases gradually throughout life following the peak in the early years (Figure 5.4). An average 60-year-old male has a max $\dot{V}_{O_2}$ of 2.2 l/min (Åstrand, 1969; Robinson, 1938). This translates into 134 W of external work, which represents a 17% (23/134) workload for the lathing task. Thus, at age 60, the relative load imposed by the lathing task has increased by a factor of 1.7 compared to that at age 18. Often, however, an increase in job skills makes the job less energy requiring. This might compensate for the decreased work capacity.

The average woman has a lower work capacity than the average

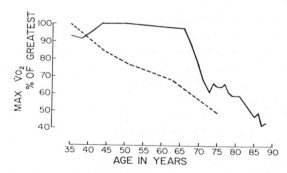

Figure 5.4 The maximum oxygen uptake decreases with age. The figure shows results for two individuals. The rate of decrease varies highly among individuals.

man. For the same task she would operate at a higher relative work load. At age 18, her maximal aerobic capacity might be 2.6 l/min (876 W) and at age 60 about 1.8 l/min (606 W; Åstrand, 1969). This translates to maximum external work capacities of 167 and 115 W, respectively. Her relative workloads on the lathing task would be 14% at age 18 and 20% at age 60. If the woman selected for the lathing job were matched with the man in terms of height, weight, and physical fitness, in terms of maximum aerobic capacity, the woman would be at no disadvantage in comparison to the man. The average woman, however, is 10 cm shorter and weighs 13 kg less than the average man (Laubach, 1976).

The efficiency with which physical work is performed and the ability to sustain a high work rate depend greatly on the state of an individual's health and level of nutrition. A few analyses have been applied to the study of labor-intensive economies with respect to the relationship between malnutrition and productivity. Spurr et al. (1978) suggested that malnutrition in young children may result in depressed work capacities as adults and further exacerbate the depressed productivity of populations in developing countries. This topic certainly deserves further study.

5.4 EFFECTS OF THE TASK ON WORK PERFORMANCE

Various factors associated with physical tasks affect task performance and productivity. The factors include the physical difficulty, or work rate; the task duration; and the work posture required by the task.

5.4.1   TASK DIFFICULTY

Table 5.2 shows the energy costs for several construction tasks. The figures in the table might be slightly misleading, since individual differences in skill, motivation, body weight, age, sex, involvement of different muscle groups, general health, and degree of physical training all affect the energy expenditure. Accurate determinations of energy costs can be made only for individuals performing specific tasks.

5.4.2   TASK DURATION

It was shown earlier that the relative workload depends both on the job work rate requirement and an individual's maximum aerobic ca-

TABLE 5.2   Average Values of Energy Expenditure for Various Tasks performed by Young Adults[a]

Activity	males (kcal/min)	Activity	Females (kcal/min)
Rest		*Rest*	
Sleeping	1.1	Sleeping	0.9
Sitting	1.4	Sitting	1.1
Standing	1.7	Standing	1.4
Work		*Work*	
Office work	1.8	Office work	1.6
Walking, level	2.6	Floor sweeping	2.0
Walking, carrying 10 kg	4.0	Walking, level	2.2
Truck driving	1.6	Light industry	3.4
Bricklaying	2.5	Walking, carrying 10 kg	3.4
Carpentry	4.1		
Sawing, power saw	4.8	Scrubbing floors	4.0
Pushing wheelbarrow	5.0		
Average construction work (laborers)	6.0		
Shoveling	6.2–10.0		
Pulling carts	6.8–10.8		
Sledge hammering	6.0-9.0		
Digging, spade	8.4		
Sawing	9.0		
Felling trees	9.8		

[a] To convert values to watts, multiply by 70.

pacity. Tasks that demand a high percentage of the maximum aerobic capacity cannot be sustained for as long as tasks that require a smaller proportion of that capacity.

Michael et al. (1961) have shown that a person of average physical fitness can work continuously (50 min each hour) for 8 hr at 35% of max $\dot{V}_{O_2}$ without producing undue fatigue. Figure 5.5 shows the length of time that physically trained, average, and untrained persons can work at a task, depending on how taxing the work is. The lathing job cited earlier required, at most, 20% of maximal aerobic capacity. The curves in Figure 5.5 predict that this task can be carried on throughout an 8-hr work day.

5.4.3 WORK POSTURE

A work posture used consistently by a worker at his job may produce transient or long-term changes in the musculoskeletal system. The curvature of the upper spine, for example, will increase through the years for a person who works at a table that is too low. Many of the complaints by workers at industrial medical centers are related to muscle pain that is due to strain induced by work postures (Table 5.3). It is thought that postural pain can place as effective a limit on productivity as does physical work capacity (Corlett, 1979).

There are several general principles for designing a task so that work posture does not inflict pain. Corlett (1979) presented a list of

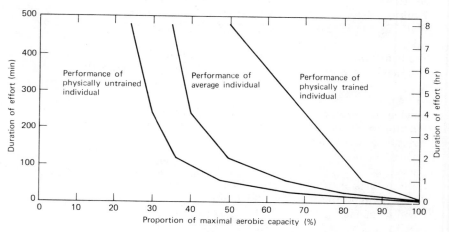

Figure 5.5 The greatest amounts of time that physically trained, average, and untrained individuals can work based on the relative physical difficulty of the work task.

TABLE 5.3 Postures and Related Complaints of Pain[a]

Posture	Complaint of postural pain
Standing	Feet, lower back
Sitting without lower back support	Lower back
Sitting without back support	Entire central back
Sitting without proper footrests	Knees, legs, lower back
Sitting with elbows on high surface	Upper back, lower neck
Unsupported arms or arms reaching upward	Shoulders, upper arms
Head bent back	Neck
Trunk bent forward (stooping)	Lower back, entire central back
Cramped position	Involved muscles
Maintenance of joint in extreme position	Involved joint

[a] (From van Wely, 1970).

important principles. (Rules for body motions were also listed by Corlett. Essentially, these principles stress Gilbreth's and Barnes' emphases on the correct uses of body momentum and gravity, the placement of tools, and the need for rhythmic and symmetrical motions to minimize fatigue and maximize productivity.) Rules for work posture include the following:

1. The worker should be able to maintain an upright and forward-facing posture during work.
2. The task should be visible with the trunk upright and the head either upright or inclined slightly forward.
3. The back should be supported during sitting work. The weight of the body should be supported equally on both feet during standing work. Ideally, the task should permit the worker to adopt any of several equally safe and healthy postures without reducing productivity.
4. Work should be performed at a height below shoulder level. Even occasional exertion of force above the shoulder level should be avoided. Where light hand work must be performed above the shoulder level, upper arm rests must be available.
5. Work should be performed with the joints at about the midpoint of their range of movement. This is particularly important for the joints of the neck, trunk, and arms.

6. Muscular force should be exerted so that large (groups of) muscles can be used. The direction of motion should be parallel with the involved limbs.

7. When the repeated use of muscular force is required, the task should be designed so that the worker can use either arm or either leg.

Several different methods may be used to analyze work posture. They include observation, photography, chronocyclography, and computer-aided posture analysis.

The Ovako Working Posture Analysis System (OWAS) was designed for use by work-study engineers (Karhu et al., 1977). This system serves as an example of how the postural requirements of a task may be analyzed. The system combines physical descriptions of postures with worker ratings of postural comfort. The descriptors of posture used in the OWAS are based on the positions of the back, arms, and legs, allowing a given posture to be described as a three-digit code. The digits are defined as follows:

Back (First Digit)

1. Straight.
2. Bent.
3. Straight and twisted.
4. Bent and twisted.

Arms (Second Digit)

1. Both arms at or below shoulder level.
2. One arm at or above shoulder level.
3. Both arms above shoulder level.

Legs (Third Digit)

1. Weight on both legs, straight.
2. Weight on one leg, straight.
3. Weight on both legs, bent.
4. Weight on one leg, bent.
5. Weight on one leg, kneeling.
6. Body being moved by the legs (walking).
7. Both legs hanging free (sitting).

The posture of an individual to toenail a stud into a plate would be described as 2–1–5. The posture would be 1–1–5 if the individual maintained a straight back while leaning forward from the waist to do the nailing. The posture of an individual carrying a sack of cement on one shoulder, steadying it with one hand, would be 1–2–6.

Workers were asked about the comfort of the various postures. A four-point rating scale was used: extremes of the scale were "normal posture with no discomfort and no ill effect on health" and "extremely bad posture, short exposure leads to discomfort, ill effect on health possible." This rating was combined with the posture description to classify work postures in four different categories:

Class 1. Normal postures that do not need any special attention except in some special cases.

Class 2. Postures that must be considered during the next regular check of work methods.

Class 3. Postures that need consideration in the near future.

Class 4. Postures that need immediate attention.

The OWAS has been used with success in industry. In particular, the system succeeded where others had failed, in providing a solution to some problems associated with bricklaying.

A technique for posture analysis developed by the Department of Engineering Production of the University of Birmingham (England) provides an alternative that is much more detailed (Corlett, 1979). Like the OWAS, this system uses observation and worker reports of postural comfort. Unfortunately, the availability of reliable, objective, health-related data dealing with work posture is quite limited. These data are difficult to acquire since research programs are rarely established in industry to collect the necessary information. As a substitute, techniques using subjective analyses of work posture, like the ones described above, can be used. These techniques actually provide very accurate data, much more than most people would believe at first sight.

5.5 BACK PROBLEMS

Some incidence of back pain or injury appears to be inevitable in any industry and is a manifestation of the normal aging process. The vertebral disks, which function as shock absorbers between vertebrae, slowly begin to degenerate at about 25 years of age. By age 60 to 65, nearly all of the disks have degenerated. Many individuals with disk degeneration will also experience back pain. The pain is intermittent and can be brought about by mild reaching, turning, or lifting. The pain is relieved primarily by bed rest and other mild treatments, such as warm baths and aspirin, as prescribed by a physician (Adams et al., 1977). Back pain problems increase with age.

The predominant back injury, disk herniation, becomes more likely as disk degeneration progresses with age and the structural relationships of the spinal column change. The final trigger for herniation is often a rise in pressure inside the chest, abdomen, and spinal disks brought about by sneezing or lifting. The pressure of displaced disk material against a spinal nerve leads to inflammation and pain.

Construction workers have more back injuries than workers in most other occupations. The National Safety Council reported statistics on workers' back injuries in California in 1975 and in Pennsylvania in 1977 (Table 5.4). Physically active workers are seen to suffer greater proportions of back injuries. Construction workers ranked second in California and third in Pennsylvania among occupational groups for which back injuries were reported. The result of back injury and of low back pain, an associated problem, is lost work time and a high level of demand for medical treatment (Adams et al., 1977).

Special physical training can be useful for preventing or minimizing back pain. The muscles of the trunk are used to counterbalance the forces on the spinal cord during lifting. Exercises such as running and

TABLE 5.4   Risk of back Injury Related to Occupation[a]

Occupation	California			Pennsylvania		
	% of workers	% of back injuries	Relative risk	% of workers	% of back injuries	Relative risk
Mining	0.4	0.7	1.75	1.1	4.8	4.36
Construction	3.7	7.1	1.92	4.5	6.3	1.40
Manufacturing	20.1	26.4	1.31	29.4	38.6	1.31
Transportation and public utilities	6.1	10.5	1.72	5.6	11.2	2.00
Wholesale and retail trade	22.8	21.1	0.92	20.1	16.3	0.81
Finance, insurance, real estate	5.8	1.8	0.31	4.8	0.9	0.18
Services	20.4	14.3	0.70	18.7	10.8	0.58
Government	20.7	18.1	0.87	15.8	11.1	0.70

[a] (*Accident Facts*, National Safety Council, 1978).

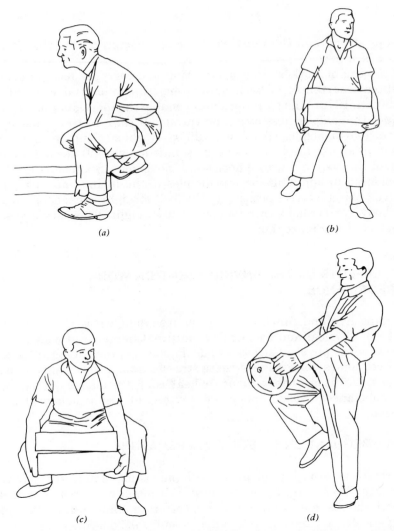

Figure 5.6 Illustration of correct lifting techniques. (*a*) Illustrates a person squatting while lifting (principle 1). The angle of the knee of the front leg is approximately 90°. The arms are held close to the body and the back is straight. Before raising the load he will tuck his chin in. This tends to further straighten the back while lifting. After raising the load he is immediately ready to move horizontally by using the momentum of the body weight (principle 2). (*b*) A weight should be carried with arms straight. This reduces the tension in the upper arm and shoulder muscles. (*c*) The arms should remain straight while lifting. The feet are placed apart to prepare for forward movement. (*d*) This illustrates how a load can be raised to bench height by using the leg muscles. This practice reduces the risk of back strain. (Adapted from *Kinetic Methods of Manual Handling in Industry* (Occup. Safety Health Ser. No. 10). Copyright *1972*, International Labour Organisation, Geneva).

sit-ups strengthen the muscles of the trunk, minimizing the chances for back injury.

Training of workers in proper lifting and carrying techniques also reduces occurrences of back pain (American Industrial Hygiene Association, 1970). The training should enable the individual to use techniques that impose less strain on the spinal cord. Lifting properly involves two principles (Himbury, 1967): (1) when lifting an object from the ground, use the leg muscles by squatting; and (2) use the momentum of the body to initiate horizontal movement. Six factors are to be considered in applying these principles to the lifting and carrying of a given object: correct position of the feet; straight back; arms close to the body; correct hold; chin in; use of body weight. These principles are illustrated in Figure 5.6.

5.6 EFFECTS OF THE ENVIRONMENT ON WORK PERFORMANCE

Humans live and function in an environment with several stresses that are physical and chemical in nature. Chapter 3 provided an overview of the physical stresses and Chapter 4 of the chemical stresses. This section reviews some of these from a work capacity point of view. We discuss the influence of heat, cold, altitude, and some chemical substances (carbon monoxide and ozone) on physiological mechanisms.

5.6.1 REGULATION OF BODY TEMPERATURE

There are several physiological mechanisms for regulating body temperature. These are under involuntary control by nerve cells in the hypothalamus, a structure in the lower brain, and maintain the body temperature within a narrow range (about ± 0.5 degrees around 37°C). This process is known as "thermoregulation." Body temperature exhibits daily variations: it peaks in the late afternoon and reaches its lowest level in the early morning.

In order that body temperature be kept within the narrow, regulated range mentioned above, the amounts of heat gained and lost by the body over a short span of time must be equivalent. There are three ways in which a human can adjust to a hot or a cold environment: adaptation, acclimation, and acclimatization. These mechanisms can compensate rather effectively for exposure to heat. However, they are not so effective in compensating for a cold environment. The unclothed human is unable to maintain a stable body temperature when the ambient temperature drops to 26°C or below. Adaptation implies a genetic change as a result of natural selection. Acclimation refers to physiological changes in response to ambient temperature changes (for example, sweating or shivering). Acclimatization refers to more enduring changes in physiological mechanisms that enable an individual to work in an extreme environment such as the tropics (Horvath, 1979).

Heat is exchanged between the body and the environment by means of four physical processes: conduction, convection, radiation, and evaporation.

Conduction is the process whereby heat is transferred by body contact with a solid object. It depends on the difference in temperature between the body and the object. Heat transfer by convection refers to the temperature exchange produced by moving air. Convection is a more important source of heat exchange than conduction. The amount of convection increases with the magnitude of the difference between skin and air temperature. A significant portion of "sensible" heat transfer between skin and air is by combined conduction–convention. Transfer by conduction–convection also occurs in the respiratory tract.

The body exchanges heat with surrounding objects by infrared radiation. Radiated heat is exchanged between bodies of different temperatures. The surface of the human body and clothing emits radiation only in the long infrared portion of the light spectrum. Surrounding objects generally emit at long infrared as well as higher frequencies.

Evaporative heat loss occurs at the body surface. Moisture is present on the skin because of diffusion from deeper tissue (insensible perspiration) and/or because of sweat gland activity (sensible perspiration). When the moisture evaporates, heat is taken from the body surface. The evaporation is a function of air speed and the difference in vapor pressure between the sweat (at skin temperature) and the air. In hot, moist environments, evaporative heat loss is limited by the capacity of the ambient air to accept additional moisture. This limits the cooling of the body. In a hot, dry environment, the evaporative heat loss is limited by the amount of perspiration that can be produced by the worker. The maximum sweat production that can be maintained by the average man throughout the day is 1 l/hr, corresponding to a heat loss of 600 kcal/hr.

5.6.2   EXPOSURE TO HEAT

There are several measures that can be used to assess heat stress. The complexity of heat transfer mechanisms has made it difficult to develop a measure that is useful in all circumstances. The National Institute of Occupational Safety and Health has concluded that wet bulb globe temperature (WBGT) is, at present, the most useful industrial measure of heat stress in workers; the design of a WBGT measurement device is given (NIOSH, 1972).

The WBGT measure takes into account radiant heat, air temperature, air velocity, and air humidity, WBGT is calculated from measurements of globe temperature (T_G), dry-bulb temperature (T_D), and wet-bulb temperature (T_W). It is measured differently in shade than in sunlight:

$$\text{WBGT}_{\text{shade}} = 0.7\,T_W + 0.3\,T_G$$
$$\text{WBGT}_{\text{sunlight}} = 0.7\,T_W + 0.2\,T_G + 0.1\,T_D$$

Figure 5.7 shows tentative maximum values of heat stress exposure.

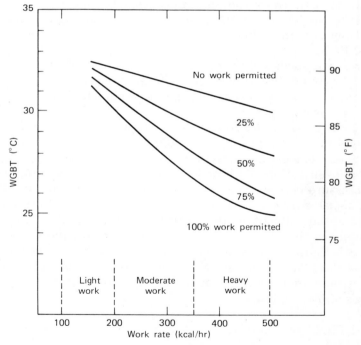

Figure 5.7 Permissible work rate as a function of heat exposure. WGBT is given in degrees Fahrenheit and Celsius. At the bottom of the figure, light, moderate, and heavy work are defined in terms of kilocalories per hour.

In the figure, WBGT is measured in degrees Celsius or Fahrenheit. Exposure limit values for light, moderate, and heavy work are given for different work rates.

Heat loss is regulated by the hypothalamus. In a hot environment, blood flow to the skin increases, resulting in a higher skin temperature. The higher skin temperature causes conductive–convective, evaporative, and radiative heat losses that cool the blood. The cooler blood then circulates through the body, accepting more internal heat for transfer to the surface of the body. Heart rate increases in order to meet the additional blood flow requirement.

The water and salt balance of the body and the capacity of the sweat glands determine the amount of sweat produced. Observations of industrial workers show that they can produce at least 1 l of sweat per hour during an 8-hr workday. Many times, the worker does not realize the extent of water and salt loss. Thus during a long period of hard work, the individual may become dehydrated without realizing it.

Wetting of the clothes worn by individuals working in a hot environment increases the evaporative heat loss since the water evaporates in the same way as perspiration. The degree to which clothing can be wetted influences the capability to maintain thermal equilibrium.

There are several symptoms of water and salt depletion (Adolph, 1947; Marriott, 1950). A water deficit of approximately 2% of body weight leads to thirst, decreased urine output, irritability, and aggressiveness. A water deficit in excess of 6% results in impairment of physical and mental performance. Salt depletion also produces serious problems. A deficit of 0.5 g/kg of body weight results in giddiness, mild muscle cramps, and, eventually, fainting. A progressive decrease in body weight over a period of days can warn a worker of a chronic salt deficit. Provision of water and salt at the workplace can prevent these problems; this is regulated in the OSHA standards for construction work.

Repeated exposure to hot environments leads to an improved tolerance to the heat load. However, individuals show marked differences in their tolerances and some individuals cannot acclimatize adequately. During acclimatization to heat, there are progressive decreases in rectal temperature, skin temperature, and working heart rate, and increased sweat rate (Robinson et al., 1943; Nelson et al., 1947). These processes can be completed in 1 to 10 days of exposure to a hot environment.

Mere exposure to heat without work confers little acclimatization. The time required for acclimatization is reduced when people perform physical work in the heat. A study by Horvath and Shelley (1946)

shows that working for 1 hr in a hot environment produced acclimatization which enabled work for an additional 4 hr.

Age does not appear to influence acclimatization in men. Robinson
et al. (1965) restudied men (44 to 60 years of age) who had been evaluated 21 years earlier for their responses to heat exposure. These older
individuals acclimatized as well as they had done when they were
younger.

Older women do not tolerate heat stress as well as younger women.
One study showed 10 elderly women had higher core temperatures and
a lower mean skin temperature than young women. At what point in
life these differences become apparent and why they occur is not known
(Cleland et al., 1969).

Acclimatization to a hot environment is not retained indefinitely
without exposure to heat. There can be a loss over so short a period
as a weekend. Recovery to the prior level of acclimatization will take
1 day. One week without heat exposure may require 4 days of exposure
to reattain acclimatization. Acclimatization is almost completely lost
after 3 to 4 weeks in a cool environment.

There are several factors that affect acclimatization. Individuals who
are engaged in hard physical work are partly heat acclimatized by
virtue of doing this work, and they require less heat exposure to develop
acclimatization. Dehydration results in a loss of acclimatization.

Few studies have been aimed at determining sex differences in reaction to heat. The research that has been done, however, has produced
the following conclusions. Women have more problems than men in
acclimatizing to heat (Burse, 1979). Females have a greater number
of sweat glands than males but the male sweat glands have greater
activity (Kawahata, 1960). The onset of sweating in response to heat
is also slower in females than in males.

There are a number of medical problems that can result from overexposure to heat: heat rash, fatigue, cramps, and heat stroke. Deaths
due to heat are common when a sudden heat wave develops and un-
acclimatized people are exposed to heat stress. The individuals most
susceptible to death are the very young and the aged, who have the
most difficulty in acclimatizing to heat.

5.6.3 EXPOSURE TO COLD

Several circumstantial factors are important for protection against
cold. These include moving out of the cold, building a fire, using
warmer clothing, and reducing the exposed area of the body. There are
several physiological adjustments as well. These have been studied

during short and prolonged exposures to cold. The physiological changes have been studied in three different environments: polar regions, temperature-regulated chambers, and normally occurring seasonal cold.

Several groups, such as the Australian aborigines and the Kalahari bushmen, appear to adapt to cold. Their internal temperatures can decrease while their skin temperatures remain high with no significant increase in metabolic rate (energy output). This is in contrast to Caucasians, who have high metabolic rates, unchanged or higher internal temperatures, and decreased surface temperatures under similar cold stress.

Cold tolerance can be measured in several ways: the length of time required for shivering to begin; magnitude of shivering; time until metabolic heat production increases; discomfort of the subject; and changes in circulatory activity. The responses of humans to cold are governed by thermoregulatory systems in the hypothalamus. The first protective response to cold is reduced blood flow to the skin, which reduces heat transfer from the body core to the surface. There is consequently a reduction in heat loss since the difference in temperatures between the skin and the environment is reduced. The second protective response is an increase in heat production.

Shivering is the first physiological response to cold exposure. It can involve single muscles as well as large muscle groups. Shivering originates either in the upper thorax or in the thigh. Periodic fluctuations in shivering activity are always observed. The periods of nonshivering vary from seconds to minutes.

As a result of shivering there is an increase in metabolic rate which produces additional heat to compensate for heat loss. The metabolic rate may be three to five times higher than during nonshivering periods. This increase is partly produced by voluntary body movement and nonshivering heat generation. Shivering represents the first metabolic response to cold. Nonshivering heat generation occurs as a result of prolonged exposure to cold.

Whereas increased heat production depends on metabolic processes, the heat loss from the body depends largely on the physical factors of the environment. The rate of heat loss from exposed skin is directly proportional to the temperature difference between the skin and its environment (Burton and Edholm, 1955). Relative humidity has no significant effect on heat loss but does alter the subjective sensation of cold.

Physical fitness might increase the resistance to cold. A study by Adams and Heberling (1958) showed that after a 3-week period of intensive physical training, body heat production increased. This raised the temperature of the feet and hands.

The sensitivity of the skin is diminished when skin temperature drops. Irving (1966) found that sensitivity was reduced sixfold when skin temperature dropped from 35 to 20°C. In general, numbing and loss of tactile sensitivity occur at finger temperatures of approximately 8°C, and manual performance may be decreased because of the reduction in finger and hand dexterity. This probably explains the loss of fine, coordinated movements noted to occur in individuals working in the cold.

Hand grip strength is reduced by exposure to cold. A study by Horvath and Freedman (1947) found that grip strength decreased by one-fourth even though the cold exposure was of relatively short duration. People are usually unaware of this loss of strength. There might therefore be a tendency for construction workers to overestimate their work capability in a cold environment.

Older people may have slightly different responses to cold environments. Their metabolic rates do not increase to the same degree as found in younger people. As a result, their internal temperatures might fall (Horvath et al., 1955; Watts, 1972). Older people also may be less aware that they are cold.

Women are less able to compensate for cold than men. They have lower hand, foot, and skin temperatures, lose more heat from the body surface, and are at greater risk of cold injury (Burse, 1979).

5.6.4 PRODUCTIVITY

Few empirical data exist concerning the effects of environmental temperatures on construction productivity. Data on mason productivity in the United States are available (Grimm and Wagner, 1974), and it has been estimated that for ambient temperatures outside the range from 7 to 32°C, productivity declines about 10% per 5°C, and declines more rapidly in the heat than in the cold (National Electrical Contractors' Association of America, 1974).

5.6.5 ALTITUDE

Forty million people live at altitudes over 3000 m (about 10,000 ft), and permanent settlements are found as high as 5200 m (about 17,000 ft). It may be impossible to live at altitudes above 5500 m for periods longer than a few months.

There are many scientific studies of the effects of high altitude. The earliest known studies of hypoxia (low oxygen pressure) took place in the late 1700s after Lavoisier reported the importance of oxygen in metabolism (Kellogg, 1978). Later, fatalities during balloon ascents inspired Paul Bert to publish *La Pression Barometrique* in 1878. This work reported the results of the first concerted scientific effort to determine the influence of high altitudes on man. Since then, several scientific expeditions have been conducted by mountaineers and researchers. In addition, the rapid increase of populations living at altitudes in excess of 3000 m has inspired scientific groups to study the behavioral and physiological characteristics of residents.

Ambient pressure decreases with altitude (Figure 5.8). Since the percentage of oxygen in the air remains constant (20.94%) at all altitudes, the partial pressure of oxygen is reduced with increasing altitude.

There is no question that changing one's residence from sea level to high altitude (greater than 1500 m) induces certain physiological changes. Some of these changes can result in disorders such as acute mountain sickness (AMS) and high-altitude pulmonary edema (HAPE).

Acute Mountain Sickness

Also known as Monge's disease, this disorder is rarely a problem below 2500 m. It is the most commonly observed altitude disorder. The symp-

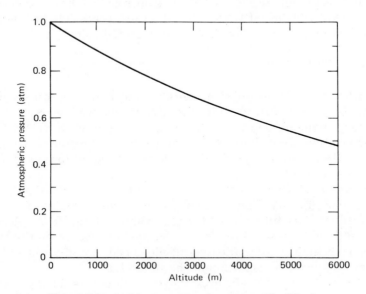

Figure 5.8 Ambient pressure decreases with altitude.

toms include headache, drowsiness, general fatigue, weakness, and difficulty with breathing on physical exertion, loss of appetite, nausea, and sleep disruption (Jarcho, 1958; Barcroft, 1925). Headache and nausea may be increased by sleep (Miller and Horvath, 1977a, b). The symptoms may persist in some individuals, which reduces their motivation and ability to work. Supplemental oxygen and the ingestion of the drug acetazolamide (Diamox) may help in reducing AMS symptoms. Generally, AMS symptoms disappear after a few days of acclimatization to altitude. Women may be less susceptible to AMS than men (Miller and Horvath, 1977a, b).

High-Altitude Pulmonary Edema

This disorder may occur in individuals who ascend rapidly from sea level to altitudes above 2500 m. It is exacerbated by physical exertion. The initial symptoms (breathing difficulties, cough, fatigue) usually appear 2 to 7 days after arrival at altitude. As the name suggests, high blood pressure in the lungs forces fluid out of the blood and into the lung tissues that separate the air from the pulmonary capillaries. Severe cases may result in death when the passage of oxygen from alveoli to blood becomes impeded. Children and teenagers are at greater risk of suffering HAPE than adults or infants (Hultgren, 1979). Indian troops moved quickly to 3300 to 5500 m from sea level have experienced a 13 to 15% incidence of HAPE (Singh et al., 1965). The incidence of HAPE in women may be lower than in men (Hultgren et al., 1961).

Effects on Performance

The ability to perform physical work is markedly reduced in direct relationship to the altitude. The capacity for physical work increases after some period of exposure but some investigators have suggested that generations (not days, weeks, or years) of continuous existence at high altitude is required before performance equal to that of native residents is possible.

The primary work-related response to altitude is a reduction in maximal aerobic capacity. Figure 5.9 shows how the maximal oxygen uptake is reduced as a function of altitude (Wagner et al., 1979). However, for any individual, the slope of this relationship depends on the length of time of acclimatization, the magnitude of body weight loss and, perhaps, on the physical condition of the individual.

The greatest reduction in max $\dot{V}_{O_2}$ usually occurs during the first few days of altitude exposure after which there may be small improve-

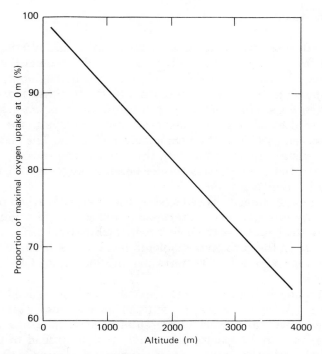

Figure 5.9   Maximal oxygen uptake decreases with altitude. The curve shows results for males with prolonged exposure (adapted from Wagner et al., 1979).

ments (Dill, 1968; Lenfant and Sullivan, 1971). High-altitude natives, on the other hand, tend to have a max $\dot{V}_{O_2}$ equivalent to that found in trained sea-level residents.

An interesting sidelight to the discussion of altitude and acclimatization is the observation that the longest-lived people are found in populations living at altitudes of about 2100 m. Georgian (Russian) and Ecuadorian villagers appear equally to share this trait. Many of these people are said to be 100 to 120 years of age.

5.6.6   AIR POLLUTION

The air surrounding the working individual contains the oxygen that is essential for the vital energy-conversion processes. But that air also contains numerous contaminants that are limiting or injurious to the individual. The internal combustion engine and a number of other sources produce at least two air contaminants that directly affect worker productivity: carbon monoxide and ozone.

Carbon Monoxide

The main effect of carbon monoxide (CO) is that it inhibits the transfer
of oxygen from the lungs to the rest of the body. This is the CO poisoning effect. Carbon monoxide combines with hemoglobin, the oxy-
gen-transporting red pigment in blood, about 210 times more readily
than oxygen does. Unfortunately, the body's oxygen sensors, located
in the arterial blood near the heart, are not sensitive to this replacement of oxygen by CO on the hemoglobin molecules. The body cannot
use CO to sustain fuel combustion; thus in the presence of relatively
low concentrations of CO, body tissues become oxygen-starved and are
hampered in their functions.

Maximal work capacity is reduced by the effects of carbon monoxide
poisoning. Apparently, maximal capacity effects are first noted when
CO binds with about 5% of the body's hemoglobin (Horvath et al., 1975).
This increase in the carboxyhemoglobin level can be brought about by
breathing approximately 100 parts per million (ppm) CO in the air
during work.

The reduction in maximal work capacity by CO may lead to poorer
productivity. As noted above, a reduction in maximal work capacity
causes the individual to work at a greater relative load on a given job
than when maximal capacity is normal. This may lead to the early
onset of fatigue in the worker or to a reduction in work rate.

Ozone

As was discussed in Chapter 3, ozone (O_3) is produced by welding.
Ozone reduces lung ventilation during physical work. The effects of
O_3 on ventilation are such that maximum voluntary lung inflation and
maximum expiratory flow rate are reduced after work (Folinsbee et
al., 1978). In other words, the breathing rate and breathing volume
of the worker are reduced by O_3. During light physical work, these
effects become noticeable when the O_3 concentration is 0.50 ppm and,
during moderate to heavy work, there are effects at 0.30 ppm O_3.

The effect of these responses on work capacity has been quantified.
The maximal work capacity of subjects exposed to 0.75 ppm O_3 (a typical peak level for a smoggy day in Los Angeles) was reduced by about
10% (Folinsbee et al., 1977).

5.7 CONCLUSION

There are many factors that influence man's ability to successfully
perform physical work in the present-day work environment. They

include age, sex, health, nutritional status, physical fitness, heat, cold, altitude, air pollution, and very likely others. The effects of most of these factors are different in developing countries than they are in the highly industrialized nations, which have generated most of the available data concerning work physiology. It is important that the techniques developed to evaluate work capacity be used to consider the relative cost of work in environments and conditions that have not yet been investigated. For example, it may be more efficient to clear a roadway with a bulldozer than with manual labor under one set of conditions, but not under another set. Physiological costs of work must be considered along with economic, social, and psychological costs.

REFERENCES

Adams, R. L., Cauthen, D. B., Neidre, A., and Washburn, K. B. Low Back Pain. *Current Prescribing,* **11**, 60–70, 1977.

Adams, T. and Heberling, E. J. Human Physiological Responses to a Standardized cold Stress as Modified by Physical Fitness. *Journal of Applied Physiology,* **13**, 226–230, 1958.

Adolph, E. F. *Physiology of Man in the Desert.* New York: Wiley, 1947.

Altman, P. L. and Dittmer, D. S. (Eds.). *Metabolism.* Bethesda, Md.: Federation of American Societies for Experimental Biology, 1968.

American Industrial Hygiene Association. Ergonomics Guide to Manual Lifting. *American Industrial Hygiene Association Journal,* **31**, 511–516, 1970.

Åstrand, I. Aerobic Work Capacity in Men and Women with Special Reference to Age. *Acta Physiologica Scandinavica,* **49** (Suppl. 169), 1969.

Åstrand, P.-O. and Rodahl, K. *Textbook of Work Physiology.* New York: McGraw-Hill, 1970.

Barcroft, J. Respiratory Functions of the Blood. In *Lessons from High Altitude,* Vol. I. London: Cambridge University Press, 1925.

Brown, A. C. and Brengelmann, G. Energy Metabolism. In T. C. Ruch and H. D. Patton (Eds.), *Physiology and Biophysics.* Philadelphia: Saunders, 1965.

Burse, R. L. Sex Differences in Human Thermoregulatory Response to Heat and Cold Stress. *Human Factors,* **21**, 687–699, 1979.

Burton, A. C. and Edholm, O. G. *Man in a Cold Environment.* London: Edward Arnold, 1955.

Cleland, T. S., Horvath, S. M., and Phillips, M. Acclimatization of Women to Heat after Training. *Internationale Zeitschrift für Angewandte Physiologie,* **27**, 15–24, 1969.

Corlett, E. H. *Posture: Its Measurement and Effects.* Birmingham, England: University of Birmingham Dept. of Engineering Production, 1979.

Dill, D. B. Physiological Adjustments to Altitude Changes. *Journal of the American Medical Association,* **205**, 123–129, 1968.

Folinsbee, L. J., Drinkwater, B. L., Bedi, J. F., and Horvath, S. M. The Influence of Exercise on the Pulmonary Function Changes Due to Exposure to Low Concentrations of Ozone. In L. J. Folinsbee, J. A. Wagner, J. F. Borgia, B. L. Drinkwater, J.

A. Gliner, and J. F. Bedi, *Environmental Stress: Individual Human Adaptations.* New York: Academic, 1978.

Folinsbee, L. J., Silverman, F., and Shephard, R. Decrease in Maximum Work Performance Following Exposure to Ozone. *Journal of Applied Physiology,* 42, 531–536, 1977.

Grimm, C. T., and Wagner, N. K. Weather Effects on Mason Productivity. *Proceedings of the American Society of Civil Engineers,* 100, 319–335, 1974.

Himbury, S. *Kinetic Methods of Manual Handling in Industry.* Geneva: International Labour Office, 1967.

Horvath, S. M. Evaluation of Exposures to Hot and Cold Environments. In L. V. Cralley and L. J. Cralley (Eds.), *Patty's Industrial Hygiene and Toxicology,* Vol. 3, *Theory and Rationale of Industrial Hygiene Practice.* New York: Wiley, 1979.

Horvath, S. M., and Freedman, A. The Influence of Cold upon the Efficiency of Man. *Journal of Aviation Medicine,* 18, 158–164, 1947.

Horvath, S. M., Radcliffe, C. E., Hutt, B. K., and Spurr, G. B. Metabolic Responses of Old People to a Cold Environment. *Journal of Applied Physiology,* 8, 145–148, 1955.

Horvath, S. M., and Shelley, W. B. Acclimatization to Extreme Heat and Its Effect on the Ability to Work in Less Severe Environments. *American Journal of Physiology,* 146, 336–343, 1946.

Horvath, S. M., Raven, P. B., Dahms, T. E., and Gray, D. J. Maximal Aerobic Capacity at Different Levels of Carboxyhemoglobin. *Journal of Applied Physiology,* 38, 300–303, 1975.

Hultgren, H. N. High Altitude Medical Problems. *Western Journal of Medicine,* 131, 8–23, 1979.

Hultgren, H., Spickard, W., and Hellriegel, K. High Altitude Pulmonary Edema. *Medicine,* 40, 289–313, 1961.

International Commission on Radiological Protection. *Report of the Task Group on Reference Man.* Oxford: Pergamon, 1975.

Irving, L. Adaptations to Cold. *Scientific American,* 214(1), 94–101, 1966.

Jarcho, S. Mountain Sickness as Described by Fray Joseph de Acosta, 1589. *American Journal of Cardiology,* 2, 246–247, 1958.

Karhu, O., Kansi, P., and Kuorinka, I. Correcting Working Postures in Industry: A Practical Method for Analysis. *Applied Ergonomics,* 8, 199–201, 1977.

Kawahata, A. Sex Differences in Sweating. In H. Yoshimura, K. Ogata, and S. Itoh (Eds.), *Essential Problems in Climatic Physiology.* Kyoto: Nankado Publishing, 1960, pp. 169–184.

Kellogg, R. H. Some High Points in High Altitude Physiology. In L. J. Folinsbee et al. (Eds.), *Environmental Stress: Individual Human Adaptations.* New York: Academic Press, 1978, pp. 317–324.

Lange-Andersen, K., Masironi, R., Rutenfranz, J., and Seliger, V. *Habitual Physical Activity and Health.* Copenhagen: World Health Organization, 1978.

Laubach, L. L. Comparative Muscular Strength of Men and Women: A Review of the Literature. *Aviation Space and Environmental Medicine,* 47, 534–542, 1976.

Lenfant, C., and Sullivan, K. Adaptation to High Altitude. *New England Journal of Medicine,* 284, 1298–1309, 1971.

Maron, M. B., Horvath, S. M., Wilkerson, J. E., and Gliner, J. A. Oxygen Uptake Measurements During Competitive Marathon Running. *Journal of Applied Physiology,* 40, 836–838, 1976.

Marriott, H. L. *Water and Salt Depletion*. Springfield, Ill.: Thomas, 1950.

Michael, E. D., Hutton, K. E., and Horvath, S. M. Cardiorespiratory Responses During Prolonged Exercise. *Journal of Applied Physiology,* **16,** 997–1000, 1961.

Miller, J. C., and Helander, M. H. The 24-Hour Cycle and Nocturnal Depression of Human Cardiac Output. *Aviation Space and Environmental Medicine,* **50**(11), 1139–1144, 1979.

Miller, J. C., and Horvath, S. M. Cardiac Output During Human Sleep. *Aviation Space and Environmental Medicine,* **47**(10), 1046–1051, 1976.

Miller, J. C., and Horvath, S. M. Cardiac Output During Sleep at Altitude. *Aviation Space and Environmental Medicine,* **48**(7), 621–624, 1977a.

Miller, J. C., and Horvath, S. M. Sleep at Altitude. *Aviation Space and Environmental Medicine,* **48,** 615–620, 1977b.

National Institute of Occupational Safety and Health, *Criteria for Recommended Standard: Occupational Exposure to Hot Environments* (HSM 72-10269). Washington, D.C.: U.S. Department of Health, Education and Welfare, 1972.

National Electrical Contractors Association of America, *The Effect of Temperature on Productivity*. Washington, D.C.: 1974.

Nelson, N., Eichna, L. W., Horvath, S. M., Shelley, W. B., and Hatch, T. F. Thermal Exchanges of Man at High Temperatures. *American Journal of Physiology,* **151,** 626–652, 1947.

Robinson, S. Experimental Studies of Physical Fitness in Relation to Age. *Arbeitsphysiologie,* **10,** 251, 1938.

Robinson, S., Belding, H. S., Consolazio, F. C., Horvath, S. M., and Turrell, E. S. Acclimatization of Older Men to Work in Heat. *Journal of Applied Physiology,* **20,** 583–586, 1965.

Robinson, S., Turrell, E. S., Belding, H. S., and Horvath, S. M. Rapid Acclimatization to Work in Hot Environments. *American Journal of Physiology,* **140,** 168–176, 1943.

Singh, I., Kapila, C., Khauna, P. K., Nanda, R. B., and Rao, B. D. P. High Altitude Pulmonary Edema. *Lancet,* **1,** 229–234, 1965.

Singleton, W. T. *Introduction to Ergonomics*. Geneva: World Health Organization, 1972.

Spurr, G. B., Barac-Nieto, M., and Maksud, M. G. Childhood Undernutrition: Implications for Adult Work Capacity and Productivity. In L. J. Folinsbee et al. (Eds.), *Environmental Stress: Individual Human Adaptations*. New York: Academic Press, 1978.

Wagner, J. A., Miles, D. S., Horvath, S. M., and Reyburn, J. A. Maximal Work Capacity of Women During Acute Hypoxia. *Journal of Applied Physiology: Respiratory, Environmental, and Exercise Physiology,* **47,** 1223, 1227, 1979.

Watts, A. J. Hypothermia in the Aged: A Study of the Role of Cold Sensitivity. *Environmental Research,* **5,** 119–126, 1972.

CHAPTER 6

HUMAN FACTORS ENGINEERING IN CONSTRUCTION WORK

MARTIN HELANDER, Ph.D.

Canyon Research Group, Inc.
Westlake Village, California

Human factors engineering aims at modifying work procedures and machinery by taking into account the physical and psychological capabilities and limitations of human beings. There are several different names for this discipline: engineering psychology, technical psychology, and, in Europe, ergonomics. Whatever name is used, the goal is to optimize the performance, productivity, or safety of a man–machine system or work environment. This involves studying the people, tasks, and equipment involved and designing ways for these three elements to interact in an optimal way.

The first scientific efforts to study human performance in the workplace were time and motion studies conducted in the early twentieth century. However, it was not until after World War II that human factors emerged as a scientific discipline. World War II brought about the creation of complex weapon systems, new types of aircraft, and various types of electronic devices. However, there was little time available for testing the new equipment and systems, and many of them later proved to be inadequate from a human capability point of view. A frequently cited example is the task of monitoring radar or sonar displays. What first looked like an easy job—simply sitting and looking—in practice proved to be nearly impossible for human operators. In general, there was a rapid decline in the ability to detect hostile planes or submarines, for example, after the first 20 min of a watch;

this was followed by a slower but progressive decline throughout the remainder of the watch. This kind of performance deficiency (and many others) pointed out the need for studying human performance capabilities.

Since World War II, there has been considerable progress in human factors engineering in areas other than the military. Much research has been performed in traffic safety. Recently, human factors design of workplaces has received considerable attention, for instance, the design of cabs on construction machinery. There is an increasing consensus that jobs of all kinds must be designed with regard for the abilities and limitations of the people who will perform them. The consequences of not doing this can be considerable: expensive but inevitable errors will be made by workers; costly equipment may be lost in accidents; and so on. The fault lies not with the people who attempt to do their jobs but with those who make the plans for the job or design the machinery. These people must take human factors into account from the outset in planning a project or designing the machinery to be used.

The last 10 years have witnessed a much increased concern for workers' health and welfare. As a result, human engineering methodology has been applied to many industrial tasks. This is particularly true of the automotive industry, which suffered from an overemphasis on productivity aspects of the work environment through the use of time and motion studies. Many tasks in automotive and other industries must now be redesigned in order to increase worker motivation and satisfaction. The construction industry, however, has devoted hardly any attention to human factors problems (see Chapter 1).

The objective of this chapter is to provide some understanding of human factors principles and to demonstrate how they might be applied in construction work. Before we go into the applications, a model of human information processing is discussed. This points out some of the limitations and capabilities of the human brain for perceiving and processing information in the work environment. An understanding of information processing is essential for those who plan and design work tasks and environments. Following the discussion of this model, we address some subject areas of particular relevance for construction work, such as the effects of shift work on productivity and fatigue, design of illumination of the workplace, selection and training of workers, and human factors design principles as applied to construction equipment.

These subject areas were chosen because it is easy to illustrate why they are important to construction work. However, the field of human factors engineering is vast. The interested reader will find more complete information in textbooks such as those by McCormick (1976) and Kraiss and Moraal (1976).

6.1   A MODEL FOR INFORMATION PROCESSING

Any work activity imposes both physical and mental or information processing workloads. The implications of physical workload discussed in Chapter 5 are relatively well understood. The information-processing workload is equally important but unfortunately more difficult to discuss because much less is known about the brain and mental functioning than is known about the physiology of work. In this section, we present a simplified model of human information processing that can provide a framework for understanding the kind of mental activities involved in information processing.

The construction environment presents the worker with information from many different sources: machines, materials, work procedures, verbal communication, interaction with other workers, and so on (see Figure 6.1). In this environment, it can be difficult to collect information since it changes from day to day. Work crews change, different equipment is used at different stages of the job, and the building itself changes with degree of completion. Simple tasks such as finding things, communicating with other people, and looking out for safety hazards

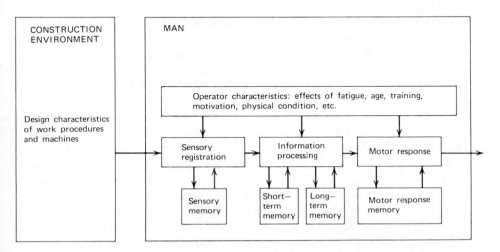

Figure 6.1   A man–environment information processing system.

can become quite complicated since it is difficult to know exactly where to look and what to look for.

Collecting and processing information at the construction site are certainly more difficult than, say, in an industrial plant where the environment remains the same. The situation may be compared with the difference between driving a car in one's hometown and in an unfamiliar city. In the unfamiliar city, most people have trouble collecting information because they do not know where to look or exactly what to look for. It is difficult to spot important sources of information (e.g., traffic signs, lights, street signs, intersections) while watching for pedestrians, bicycles, and other drivers. A driver can then be overloaded with new information, much of which may be irrelevant. That is why drivers often violate local traffic laws; an information overload makes it impossible to process critical information fast enough. Similar difficulties might be expected in a work environment that changes all the time, such as the construction site. This may result in accidents and decreased productivity.

Figure 6.1 shows that the information from the environment is first registered by the sensory system (e.g., vision, hearing, touch). Some, but not all, of the information is processed and a decision is made that results in a muscular response.

The eye and ear are the most important information receptors. The eye alone may be responsible for collecting 90% of incoming information about the environment. After it has been received by the eye or ear, the information is stored briefly in sensory memories. There it remains accessible for a fraction of a second (about the time it takes to blink). Without the sensory memories, there would not be sufficient time to extract the important information from the visual or auditory "snapshot" of the situation. The sensory memory also helps in bringing together pieces of information (the snapshots) so that the continuity of a string of events is maintained in spite of disruptions such as eye blinks.

Since humans do not have the capacity to attend to everything the sensory receptors receive, most of the information in the sensory memories remains subconscious. Only a fraction of the information in sensory storage is selected and processed consciously. This takes place in the short-term memory (STM). In the short-term memory, as the name suggests, the information fades away after a short period of time. It also has limited capacity in terms of the quantity of information it can store. In the STM, new sensory information is continuously imposed on the old and, since space is limited, almost everything is forgotten after about 30 sec (Miller, 1956).

The information in STM that is judged more important can be transferred for storage in the long-term memory (LTM). LTM can hold information for more permanent storage. If the information in STM is not important, it will be entirely forgotten unless the person decides deliberately to remember it. For example, in card playing, it is possible to keep track of the cards played by rehearsing silently. This deliberate action transfers the information from STM to LTM.

Both STM and LTM are actively involved in information processing. A simple example of how they work is found in doing multiplication. In multiplying 7×7, the LTM will automatically supply the correct answer (which was "memorized" by rehearsing in school). But when multiplying 77×77, the STM is required to remember the intermediate results until the answer is computed. This mental operation is difficult since the calculation must be executed fast enough so that the intermediate results do not fade away from the short-term memory (which "forgets" almost everything in about 30 sec).

After the information has been processed, a decision is made and the appropriate muscular response or action is triggered. For manual tasks such as braking a car, it takes anywhere from 0.5 to 2 sec to initiate the muscular response. This time lag is called "reaction time." The more complex the information, the longer the time lag between the initial sensing and the eventual action that results from information processing. For tasks that are highly familiar, many of the information processing stages are skipped and the motor response occurs more or less automatically (via reflex). This speeds up the action considerably.

Anticipated muscular actions can be stored in a motor response memory for preprogrammed execution. This is the case when there is a series of muscular responses which must be triggered in sequence. A good example is driving a car through a curve. This is largely a preprogrammed action and the necessary decision making about steering, braking, and shifting is done several seconds before entering the curve. If this were not the case, the driver would be so involved in decision making that he or she would likely drive off the road. There are numerous such preprogrammed motor activities in construction work, such as anticipating the motions of the bucket of a crane and compensating with appropriate control corrections.

6.2 SELECTIVE ATTENTION

"Selective attention" is the term used for the brain processes that sort and filter information before it ever reaches the conscious level. These processes are described below.

Humans have a limited ability to process large quantities of information. Vast amounts of information are received continuously by nerve endings all over the body. Of this input, only a small fraction can be processed in the brain. Selective attention involves an automatic sorting process of the information; only the most important or relevant information gets through for conscious consideration. There are mechanisms in the information-processing system that automatically focus attention on stimuli with particular attention evoking characteristics such as novelty, relevance, complexity, and prominence.

The brain's "automatic focus" leads to an enormous reduction in the amount of information from the time it is received by sensory nerve endings to its final storage in the LTM. Table 6.1 shows that the amount of information is reduced so that less than one-billionth of what was received can be stored in LTM. Since there is such a tremendous reduction of information, it is clear that man cannot always be a good monitor of the surrounding world. However, the system of selective attention is very efficient in reducing information so that only the most "important" remains. For example, it is not necessary to remain continuously aware of elements of the environment that are not changing or presenting problems. If, for instance, you put on a hardhat in the morning, it is not important to remain aware that you are wearing it. The sensory receptors in your scalp may continue to convey this information, yet it would be distracting to be constantly aware of the hat. Selective attention mechanisms focus the attention on more important information being received at other nerve endings. Your hardhat might come up for conscious consideration again only if you bumped your head on something.

TABLE 6.1   Reduction of Information During the Various Stages in Information processing[a]

Process	Maximum flow of information (bits/sec)[b]
Sensory reception	7,000,000,000
Nerve connections	3,000,000
Conscious information processing	16
Long-term memory storage	0.7

[a] After Steinbuch (1962).
[b] This is a computer analogy. One bit is a yes–no type of information. In this case it indicates the number of nerve impulses.

When the selective mechanisms are faced with too much novel information, confusion and overloading occur. This is often the case in unfamiliar situations. The presorting mechanisms of selective attention are not able to reduce adequately the information since they don't know what is most important. The result is a jumble of impressions that must be consciously sorted (a relatively slow process) and may not be understood in time to avert problems. In a work environment, this can lead to errors or accidents as a worker attempts to perform a job or simply walk across the work site without an accurate assessment of possible hazards.

Selective attention can be controlled voluntarily as well as automatically. Imagine yourself at a crowded, noisy party. What conversation will you join? You can listen to a conversation behind you, or on the right or left. It is possible to select which of the many conversations you will attend to, but it is not possible to take part in two or more different conversations at the same time. Several conversations can be followed by listening to a few words from each and keeping track of who is where. But if the talk is at all serious or thoughtful, then the sense of each conversation is lost when you try to attend to several (Lindsay and Norman, 1972).

The same problem is often present at a construction site where there are numerous competing sources of information. Paying attention to instructions or a conversation may reduce the attention to other important inputs, such as the warning signals from a moving crane or a backing truck.

Attention is something of a two-edged sword. On the one hand it allows us to follow one set of events in spite of a background of competing information. Without this selective ability, life would be chaotic, since we could make no sense out of what happens in the world. On the other hand, selective attention limits our ability to keep track of several things at the same time. This is what happens in the case where a person sits in a meeting, concentrating hard on what to say. The contributions of the others are often lost because there is a limited ability to do two things at the same time. The consequences in a work situation are numerous. If tasks require attention to two or more things at once, chances are that sooner or later an error or accident will result. People have difficulty in attending to more than one input at a time.

Information processing capabilities differ substantially from one individual to the next, depending on personality characteristics, experience, training, age, fatigue, and other factors (see Figure 6.1). For instance, with increasing age, the sensory registration of information at the eye and ear may be impaired. In addition, neural processing may slow down so that reaction time increases. It is important to analyze how all these factors affect the capacity to work at a particular job and how the working environment can be improved to enhance information processing. This is the subject of the remainder of the chapter.

6.3 DIVISION OF WORK BETWEEN MEN AND MACHINES

Depending on the work, it is sometimes better to use a machine than a human. Machines are obviously stronger and can sometimes handle the information processing better. Given the increased automation and computerization of work, it is important to analyze exactly what types of work man should do and when machines might be chosen instead. Fitts (1951) outlined some characteristics that distinguish human performance from that of machines on a number of functions (see Table 6.2).

The list in Table 6.2 can be used to divide tasks into their components and to decide which components or subtasks should be performed by humans and which by machines. There are considerations not listed in the table, however. For example, at an assembly line, machines are used to perform most of the work. Occasionally, however, there are tasks that cannot be done easily by machine, especially those that involve complex perceptual judgments and fast evaluative responses, for instance, quality control. These are given to a human, who finds himself surrounded by machines that leave him only the odd and repetitious jobs. This usually has a substantial negative impact on work motivation and satisfaction.

6.4 SHIFT WORK

Approximately 300,000 construction workers in the United States (7% of the construction work force) are engaged in shift work (Tasto and Colligan, 1977). Shift work can be loosely defined as any arrangement of working hours other than the usual. Essentially, there are two types of shift work systems: operations that work around the clock, 24 hr a day, and those involving fewer hours than that.

People engaged in shift work receive a higher pay rate than day workers. This extra pay is considered compensation for having to work undesirable hours and being exposed to the negative effects of shift work, such as fatigue, health disorders, disruption of social life, de-

TABLE 6.2　Relative Advantages of Men and Machines

Property	Advantages and disadvantages	
	Machine	Man
Speed	Much superior	Slow speed with reaction time lag
Power	Large, unchanging power available	0.2 hp for continuous work; maximum 2 hp for 10 sec
Consistency in routine, repetitive work	Ideal	Subject to fatigue and lack of motivation
Simultaneous activities	Multi-channel	Single channel with low information processing capabilities
Computing	Fast and accurate; poor at error correction	Slow and inaccurate; good at error correction
Memory	Best for literal reproduction and short-term storage	Large store, multiple access; better for concepts and strategies
Reasoning	Good deductive; tedious to reprogram	Good inductive; easy to reprogram
Signal detection	Can sense all electromagnetic signals	Can sense only a narrow band of electromagnetic signals
Overload reliability	Sudden breakdown	Gradual breakdown
Intelligence	Incapable of switching strategies without tedious reprogramming	Can anticipate and adapt to unpredicted situations
Manipulative abilities	Only specific	Great versatility

[a] Adapted from Fitts (1951) and Singleton (1974).

creased productivity and working morale, and increased accident rates. Despite these drawbacks, many workers like shift work (Tasto et al., 1978). Any management planning to institute shift work should consider the negative effects, however, since they could drastically reduce the overall benefits. This section reviews the types of shift work systems that are usually employed, some worker health effects, and some consequences for social and family life.

6.4.1   ARRANGING FOR SHIFT WORK

A 24-hr operation usually uses the classic swing and graveyard shifts (6AM to 2 PM to 10 PM to 6 AM) on a fixed or rotating basis. For a plant that operates 5 days per week, this requires three crews. For a 7-day operation, a fourth crew is required. In both types of arrangement, it is usual for the crews to rotate by taking turns at the morning, afternoon, and night shifts.

Figure 6.2 shows some typical rotation systems for continuous shifts during a 7-day week. The numbers "2–2–2" and "2–2–3" indicate the number of days of work on each shift before a day off. The 2–2–3 system seems to be preferred by workers because of the 3-day weekend that occurs once every 4 weeks (Murrell, 1972).

There are several other ways of manning a 24-hr process. One is to operate 12-hr shifts and another is to have one crew on permanent night shift, with the day and swing shifts rotating.

	Crew	M	TU	W	TH	F	SA	SU	M	TU	
2–2–2	A	D	D	S	S	G	G	—	—		
Shift	B	S	S	G	G	—	—	D	D		
system	C	G	G	—	—	D	D	S	S		
	D	—	—	D	D	S	S	G	G	etc.	
2–2–3	A	D	D	S	S	G	G	G	—	—	
Shift	B	S	S	G	G	—	—	—	D	D	
system	C	G	G	—	—	D	D	D	S	S	
	D	—	—	D	D	S	S	S	G	G	etc.

D = Day shift, 0600–1400
S = Swing shift, 1400–2200
G = Graveyard shift, 2200–0600

Figure 6.2   Shift work patterns for continuous 2-2-2 shift and 2-2-3 shift. The crews rotate between the 4 shift patterns.

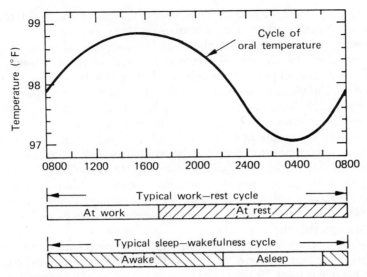

Figure 6.3 The oral temperature as well as other physiological mechanisms show a 24-hr circadian cycle.

6.4.2 HEALTH EFFECTS OF SHIFT WORK

Several of the negative effects of shift work are due to the normal differences in some body functions during night and day. Figure 6.3 illustrates how the hours around the clock can be divided into a work–rest cycle and a sleep–wakefulness cycle. These cycles determine the activity level of several physiological functions. The figure illustrates how body temperature (as an example) reaches its peak at about 4 PM and its low point at 4 AM. Many other body functions go through similar 24-hr cycles. For example, during the early morning hours, heart rate and the production of stress hormones decrease, mental activity is reduced, digestion slows down, and the secretion of urine is reduced. All these changes are conducive to a good night's sleep. At the other phase of the cycle, the reverse is true; the physiological functions are at their maximum, which is favorable for work. These physiological changes are called "circadian rhythms."

The circadian rhythms are fairly stable. For a person who starts working a graveyard shift, it will take approximately 2 weeks to adjust the circadian rhythms so that the maximum activity occurs during the night work and the low points occur during the day when the worker is sleeping. Until the 2 weeks have passed, a worker will feel tired and productivity may suffer; this decreases as time goes on.

The only way to reduce fatigue from shift work and to restore productivity is for the worker to stay on the same shift for a long period of time. However, in practice, this is rarely possible because most workers have obligations to friends and family which will conflict with a continuing nighttime job. For most workers, night shifts are therefore undesirable for more than a couple of days at a time.

There are two major health effects of shift work: lack of sleep and changes in eating habits. Persons who work late shifts sleep an average of 2 hr less per night. This is due not only to the effect of circadian rhythms but also to participation in social events and family life during the day. Eating habits are affected by lack of facilities for obtaining a hot meal during the night shifts. Although the worker's total caloric intake is usually the same, there is a tendency to eat less nutritious food, drink more coffee than normal, and use more alcohol and tobacco. Complaints about gastric and intestinal disfunction are common among shift workers. Several studies have shown that approximately 40% of shift workers have problems with loss of appetite and constipation (Tasto et al., 1978).

Shift work can also have a substantial effect on accident rates. A study by Harris and Mackie (1972) investigated 493 truck accidents, all of which had occurred as a result of the driver falling asleep or dozing while driving on interstate routes. The highest percentage of accidents occurred between 4 and 6 AM. In fact, twice as many of these accidents occurred between midnight and 8 AM than occurred during the other 16 hrs of the day (see Figure 6.4). The results are even more striking when one considers that there is much less traffic during the early morning hours than during the rest of the day. If traffic volume is taken into account (see Figure 6.4), it can be calculated that the probability of having an accident because of falling asleep at 5 AM is almost 20 times greater than the probability at noon. Clearly, much of this is attributable to the effects of the body's circadian rhythms, which make it difficult to remain alert during the early morning hours.

Some important effects of shift work are summarized below. For more complete information, the reader is referred to Rutenfranz et al. (1977).

1. Shift work has a large impact on social life. People working shifts often perceive themselves as being apart from their families and friends. There are often complaints about disturbances in family and sex life, and difficulties in maintaining relationships with friends. They encounter difficulties participating in social events scheduled for

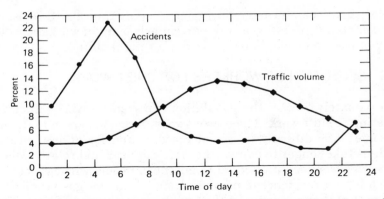

Figure 6.4 Accidents in interstate truck driving due to driver falling asleep or dozing (Harris and Mackie, 1972).

"normal" hours. As a result, shift workers are usually not active in community work or trade unions. In the case of continuous shift work, it is important to arrange as many free weekends as possible for the worker so that he can enjoy a normal social life with friends who do not do shift work. To make it easier to plan social activities, it is also important that the shift work have a regular schedule and that it not extend over a long period of time. Hence a 4-week rotating system makes it easier to plan social events than a 40-week system.

2. In order to maintain productivity, the length of a shift should be related to workload. If the work is light, the length of the shift may (with caution) be extended to 12 hr. If the work is physically heavy or imposes a high mental load on the individual, it should not exceed 8 hr.

3. Intermittent single night shifts are in most cases better than consecutive night shifts because a single night of work upsets the circadian rhythms to a lesser extent. After several night shifts, it might take a week to readjust the rhythms. A worker would then constantly suffer from fatigue, first from adjusting to night work and then from readjusting to day work. It might be argued that a permanent night shift is better since the number of readjustments of the circadian rhythms would be reduced. However, workers need some rest days during which they assume "normal" waking and sleeping hours in order to socialize with friends and family (Rutenfranz et al., 1977). Working the night shift for a long period of time would thus produce greater disruptions in the circadian rhythm than working single nights.

4. Sleep disturbances and reduction of sleeping time are the most common complaints of shift workers, particularly of night workers. At least 24 hr of free time should be allowed after night shifts.

6.4.3 SELECTION OF WORKERS FOR SHIFT WORK

Approximately 20% of the population experience severe difficulty in adjusting to shift work. The reasons vary from one person to the next: for some, it is difficult for social reasons, and for others, circadian rhythms take longer to adjust (the time required and relative discomfort varies from one individual to another). People who cannot adjust have a higher incidence of nervous problems and/or gastric disorders, and most of them drop out of shift work after a period of time. Unfortunately, it is not possible to predict with any accuracy which individuals will suffer the most. The following factors may be important and could be used in screening applicants for shift work.

1. People living alone do not adjust as easily as those with a family. This is due to the fact that the family members are generally supportive.
2. It is more difficult for people with a history of gastric or digestive disorders to adjust to shift work.
3. Epileptics have a higher incidence of fits due to reduction of sleep.
4. People with sleeping facilities that are not adequately soundproofed suffer more.
5. People over 50 years of age may be less fit for shift work.

6.5 ILLUMINATION

Vision is the most important sense for collecting the information necessary to do work. It is commonly estimated that 90% of the external information processed by the brain is visual. Therefore, it is important to design the work environment so that essential details of the work may be easily seen. This can be achieved by (1) providing adequate illumination and (2) eliminating glare sources. This is not always so simple as it might seem; in fact, there is a whole scientific discipline devoted to illumination engineering.

We first discuss how illumination is measured. We then present the current illumination standards for construction work required by the U.S. Department of Labor. In the last part of this section, we discuss two different kinds of glare and methods to avoid them.

6.5.1 MEASUREMENT UNITS

From a physical point of view, light is radiant, electromagnetic energy of a type that can be sensed by the eye. The eye is sensitive only to a narrow range of wavelengths with frequencies around 10^{15} Hz. The rest of the electromagnetic spectrum is not visible and includes, for example, X-rays, radar, and TV and radio broadcast waves.

The response of the eye to illumination involves complex photochemical reactions of the rods and cones in the retina. The cones are located in the central field of the retina and are sensitive to color. The rods are located peripherally. They do not discriminate colors but are much more sensitive to low illumination levels. The eye can adapt to a vast range of illumination levels, but it can take a long time (up to 30 min) for complete dark adaptation.

There are mixed opinions about the best way to measure illumination and several different units are currently used to measure the same or similar quantities. (Because of the many different ways in which light is quantified, there is considerable confusion among professionals.) The definitions below cover those traditionally used in the United States.

The four measures of primary importance for the measurement of light (photometry) are as follows:

1. *Luminous Intensity.* This is the power of a point source that radiates in all directions. It is measured in candelas (cd) or candle power.

2. *Luminous Flux.* This is the amount of light emitted by a point source into 1 steradian (a steradian is the unit of solid angle in space, just as a radian is a unit of angle in a flat plant). In Figure 6.5 the 1 ft^2 and 1 m^2 both subtend 1 steradian. Luminous flux is measured in lumens.

3. *Illumination (or Illuminance).* This is the amount of light reaching a surface. A common unit of measurement is footcandle (fc). In Europe, the preferred unit is lux (1 fc = 10.76 lux). The footcandle is, by definition, the illumination produced on the inner surface of a sphere of 1 ft radius when a point source of light with an intensity of one candle power is placed in the center of the sphere. Alternatively, a footcandle can be defined as the illumination on a surface of 1 ft^2 on which there is a uniformly distributed flux of 1 lumen. Illumination falls off with distance according to the inverse square law:

$$\text{illumination (fc)} = \frac{\text{luminous intensity (cd)}}{\text{distance}^2}$$

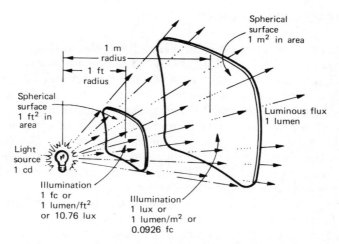

Figure 6.5 Illustration of photometric measurements. Note how the illumination is reduced to about one-tenth at 1 m compared to 1 ft. This illustrates the inverse square law.

4. *Luminance or Brightness.* This is the amount of light emitted or reflected from a surface. A common unit of measurement is the foot-lambert (fL). One footlambert is the luminance of a perfectly diffusing and reflecting surface illuminated by one footcandle. Other units for luminance are the lambert (L) and candela/m².

Table 6.3 gives a summary of the light measurement units discussed above.

6.5.2 STANDARDS FOR ILLUMINATION

Because of the adaptability of the eye, it is difficult to arrive at lighting standards that have practical value. Many standards have been published, often requiring different illumination levels for similar jobs or work situations. These standards have been subject to continuous modification; in general, the recommended illumination levels increase with each new publication. This tendency might reflect the increased concern over health and safety issues and the realization that increased illumination levels cost little to implement.

Table 6.4 lists the illumination levels required by the Department of Labor in construction areas, ramps, runways, corridors, offices, shops, and storage areas where work is in progress.

The construction standards are lower compared with other recent

TABLE 6.3 Description of Light Measurement Units

Variable	Description	Unit
Luminous intensity	Brightness of point source	Candle power, candela
Luminous flux	Flow of light	Lumen
Illumination	Amount of light reaching a surface	Footcandle
Luminance	Amount of light reflected or emitted from a surface	Footlambert

recommendations (compare Tables 6.4 and 6.5). A useful rule of thumb is that the illumination level should be 30 times higher than the level at which the task can just barely be performed. However, it is worth repeating that there are no exact lighting standards and it is usually better to apply too much light, provided glare can be avoided, if only because lighting costs are much less than labor costs (Singleton, 1972).

As people grow older, the clear substance between the lens and retina becomes increasingly clouded by small particles. These particles scat-

TABLE 6.4 Minimum Illumination Intensities for Construction Areas[a]

Illumination Intensity (fc)	Area or operation
3	General construction areas, concrete placement, excavation and waste areas, accessways, active storage areas, loading platforms, refueling, and field maintenance areas
5	Indoors: warehouses, corridors, hallways, exitways. Tunnels, shafts and general underground work areas, except that a minimum of 10 fc is required at tunnel and shaft heading during drilling, mucking, and scaling. Bureau of Mines approved cap lights are acceptable for use in tunnel heading
10	General construction plant and shops (e.g., batch plants, screening plants, mechanical and electrical equipment rooms, carpentry shops, rigging lofts and active storerooms, barracks or living quarters, locker or dressing rooms, mess halls, and indoor toilets and workrooms)
30	First aid stations, infirmaries, and offices

[a] U.S. Department of Labor (1979).

157

TABLE 6.5   Recommended Minimal Amounts of Illumination Levels

Illumination level (fc)	Type of visual task	Examples
150	Very small prolonged tasks with very small detail	Gauging very small parts; hosiery mending
100	Severe prolonged tasks with small detail	Fine assembly and machining weaving thin fibers
75	Fairly severe tasks with small detail	Drawing offices; cutting and sewing clothes
30	Ordinary tasks with medium detail	General offices; general assembly
20	Rough tasks with large detail	Stores; heavy machinery assembly
10	Casual seeing	Passages; cloakrooms

[a] Adapted from Hopkinson and Collins (1970).

ter the incoming light and reduce the contrast of visual images on the retina. Older people therefore need more illumination and particular care in eliminating glare sources in their field of view.

6.5.3   REFLECTANCE AND CONTRAST

Two other parameters of importance for describing illumination are reflectance and contrast. Reflectance is the relation between illuminance (illumination) reaching a surface and the resulting luminance of that surface. A perfectly diffusing and reflecting surface would be one that absorbs no light and scatters the illumination in the manner of a perfect matte surface. Such a surface would have a reflectance of 100%. If illuminated by 1 fc, it would have a luminance of 1 fL for all viewing angles. In actual practice, the maximum reflectance achievable for a nearly perfect diffusing (white) surface is about 95%. Reflectance is defined by the following formula:

$$\text{reflectance} (\%) = 100 \times \frac{\text{luminance (fL)}}{\text{illuminance (fc)}}$$

Contrast is the measure of luminance difference, usually between luminance of a target (L_t) and its background (L_b) as defined by the formula

$$\text{contrast} (\%) = 100 \times \frac{L_t - L_b}{L_b}$$

Contrast can be positive or negative, since the target can be brighter or dimmer than the background. The sign of contrast has no major effect on perception. In other words, a dark object against a light background can be seen as easily as a light object against a dark background at the same contrast level.

6.5.4   GLARE

there are two types of glare, disability glare and discomfort glare. As the names suggest, they have different effects on the visual system. Disability glare exists when a glare source measurably reduces a person's ability to see, whereas discomfort glare produces discomfort and fatigue but does not reduce vision directly.

Disability glare is common in situations such as driving a car at night. The headlights of an oncoming car make it difficult to see anything other than the headlights themselves.

Disability glare is not common in the ordinary work situation (except for drivers). Discomfort glare is more frequent; it occurs, for example, when a person is working facing a window. The difference in brightness between the window and the surrounding surfaces can impose a visual strain which results in headache and visual fatigue. Unfortunately, most people are unaware of the reasons for these symptoms, for discomfort glare is not one of the obvious stresses of a workplace. It should therefore be taken into account by the people who design the work environment, so that work efficiency is not impaired.

To avoid discomfort glare, the contrast between the luminance of objects in the direct visual field, the task area, and the surrounding visual field should be limited. For the center of the visual field the ratio between luminance of the brightest area and the darkest area should not be greater than 3:1. For objects located peripherally to the task, the ratio should be no greater than 10:1. For light fixtures (or windows) and the surfaces adjacent to them (such as ceilings and walls), the contrast should be no greater than 20:1. The recommended values for luminance ratios are given in Table 6.6.

To avoid glare from light sources, there are several precautions that should be taken (U.S. Government Printing Office, 1975):

TABLE 6.6   Recommended Luminance Ratios for Industrial Environments

Comparison between the brightest and darkest objects	Recommended maximum luminance ratio
In task and adjacent surroundings	3:1
In task and remote surfaces	10:1
In luminaries (windows, etc.) and surfaces adjacent to them	20:1
Anywhere within normal field view	40:1

1. Avoid bright light sources within 60° of the center of the visual field. Since most visual work is at or below the eye's horizontal position, placing light fixtures high above the work area minimizes direct glare.

2. Use indirect lighting.

3. Use several relatively dim light sources rather than a few very bright ones.

4. Use polarized light, shields, hoods, or visors to block glare in confined areas.

One way to obtain satisfactory luminance contrast levels is to paint surfaces with light colors (providing high reflectance). This also contributes to an economical distribution and use of light in a room. It is important, however, to consider that areas of high reflectance in the visual field can become sources of glare. For this and other reasons the reflectance of the various surfaces in a room generally should increase from the floor to the ceiling. Figure 6.6 illustrates the IES recommendation for reflectance values in an office. The same values apply to industrial workplaces.

6.6 DESIGN OF MACHINES

Inadequate design of construction machinery often leads to loss of productivity and worker dissatisfaction. It is therefore important to evaluate construction machines that are purchased by a construction company.

This section covers three important aspects of how equipment should be designed in order to make it easy and efficient to operate. First, we look at the problems of accommodating people of different sizes at work stations. Then we turn our attention to the problems of designing controls. Finally, some rules for designing displays are reviewed.

6.6.1 ANTHROPOMETRIC DESIGN OF THE WORK STATION

In order to fit machines and mechanical operations to construction equipment operators, it is necessary to obtain measures of the physical dimensions of the people who are going to operate the equipment. The science of measuring human dimensions is called anthropometry.

To the layman this may appear to be a straightforward process which would require relatively little research. Unfortunately, this is not the case; in fact, there are a number of complicating factors. The first problem is that the variation in sizes among individuals and among groups of individuals is usually so large that it is impossible to use average body sizes in designing work spaces or equipment. For example, if the height of a seat were designed using "average" knee height, 50% of the users (mostly women) would not be able to touch the ground.

Apart from sex differences, there are other important differences in body measures: genetic differences between ethnic groups, social class,

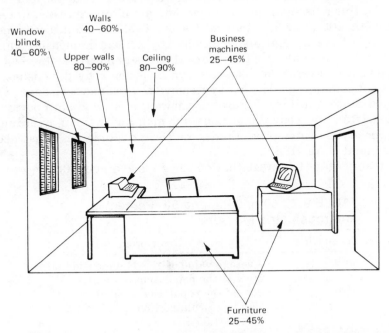

Figure 6.6 Reflectances recommended for room and furniture surfaces in offices.

and even occupation. For example, a male company director is an average of 4 in. taller than the average male unskilled worker (Singleton, 1972). Another important consideration is that there are large differences in size among people of different nationalities. For example, the average height of persons in the United States is 177 cm, in France 173, in Japan 166, and in Vietnam 160. This implies that American design standards for construction machinery (for example, truck seats and controls) would fit approximately 80% of the Frenchmen, 45% of the Japanese, and only 10% of the Vietnamese (Chapanis, 1974). Equipment built to American standards is therefore hopelessly oversized for large segments of the world's population.

It is therefore necessary to design a work space so that it can accomodate a range of individuals of different body sizes. The designer's question is, what proportion of the actual range of sizes should be considered? For example, most adults are between 5 and 6 ft. tall, but there are some men as tall as 8 ft. It would be absurd, however, to make all doorways at least 8 ft high to provide clearance for the few individuals of this height.

The standard practice in anthropometry is to design spaces to accommodate 90% of the population, disregarding the upper and lower 5%. For this reason, anthropometric data are often expressed in terms of the 5th, 50th, and 95th percentiles (see Table 6.7). Since body measurements have normal distributions, any percentile measure can be derived from the values of the mean (M) and standard deviations (S.D.).

There are several data sources available for body dimensions for different populations. The most extensive rsearch has been done by the United States military. Figure 6.7 shows the 5th and 95th percentile values (standing body dimensions) for men and women according to the Society of Automotive Engineers Design Guide for Construction Equipment (SAE, 1980).

It is important to realize that no person is average (50th percentile)

TABLE 6.7   50th, 95th, and 5th Percentile
Anthropometric Measures

Percentile	Description
50th	Average value of a body dimension (M). 50 % of the population is smaller
95th	95% of the population is smaller. Can be calculated from the formula. $M + 1.65$ S.D.
5th	5% of the population is smaller. Can be calculated from the formula. $M - 1.65$ S.D.

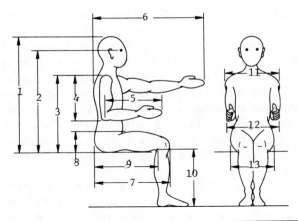

	Male			Female		
Body measurement	5th	50th	95th	5th	50th	95th
1. Sitting height	33.2	35.7	38.0	30.8	33.4	35.7
2. Eye height	29.4	31.5	33.5	27.1	29.0	31.0
3. Shoulder height	21.3	23.3	25.1	21.2	22.8	24.6
4. Shoulder to elbow	13.2	14.3	15.4	11.9	13.1	14.3
5. Forearm to hand	17.6	18.9	20.2	15.3	16.7	18.0
6. Arm reach	31.0	*a*	37.0	22.0	*a*	33.0
7. Buttock to knee	21.3	23.3	25.2	20.4	22.4	24.6
8. Thigh clearance	4.3	5.7	6.9	4.1	5.4	6.9
9. Buttock to popliteal	17.3	19.5	21.6	17.0	18.9	21.0
10. Popliteal height	15.5	17.3	19.3	14.0	15.7	17.5
11. Shoulder breadth	16.5	17.9	19.4	14.1	16.0	18.0
12. Elbow to elbow	13.7	16.5	19.9	12.3	15.1	19.3
13. Hip breadth	12.2	14.0	15.9	12.4	14.3	17.1

a Data not available.

Figure 6.7 Sitting body dimensions in inches for male and female construction workers (Source: SAE,J833a, Recommended practice, SAE, 1980).

in all body dimensions. A person of average height might be 5th percentile in sitting height and hip breadth and 95th percentile in arm reach length. Again, this emphasizes the need to design for a range of sizes rather than average sizes only.

6.6.2   USING ANTHROPOMETRIC DATA

The first step in designing a work space is to conduct an anthropometric survey or use existing data that are applicable to the group of workers. Often there are problems in dealing with this kind of information; without training in biomechanics or anthropometry, it is difficult to know how the data should be used. What, for instance, can be inferred from a "popliteal height" of 17 in.? (See Figure 6.7.) The problems of terminology can be partly overcome by using methods of design that rely less on biomechanical terminology. For example, (1) scale drawings of the work station in which scaled models of human operators may be placed, or (2) fitting trials with full-scale mock-ups (Roebuck et al., 1975). An anthropometric manikin has many uses in work station design. The manikin is a scaled-down model of the human operator, usually in two dimensions, with swivel points corresponding to the joints of the human body. These manikins are available for different percentiles, so that a range of body sizes can be tested in the proposed designs.

The other technique for gathering anthropometric data is fitting trials, using people of different sizes as models. This requires a full-scale model of the work space. Comments on the design are solicited from the individuals who test it, for example, for ability to reach controls comfortably. Usually the operator is not sitting motionless at his task, as is implied when a manikin is used to test the design. The use of fitting trials illustrates some of the problems of movement that cannot otherwise be foreseen. For example, a real person moving around in a mock-up may eventually discover a lot of ways to bump knees and shoulders on equipment that would not otherwise be evident. Likewise, the person will be able to tell the designer whether or not he or she can see two different information sources in a truck cab (e.g., a windshield and a gauge) without moving the head or releasing the controls. Although this method is more costly than other design methods, it has the advantage of providing dynamic anthropometric measures. In contrast, the method using scaled-down manikins cannot be used to obtain information on the dynamic characteristics of the work station. In fact, to obtain this kind of information with manikins is time consuming and can really never be complete.

6.6.3 CONTROLS

Controls on a machine are used to transfer operator decisions to the machine. Often the controls are regarded as being a part of the machine hardware. For design purposes, however, it is more relevant to consider them as the link between the machine and the operator. Hence the control transmits information from the man to the machine and the starting point in the design of controls must be the output characteristics of the human, for example, comfortable patterns of movement.

More often than not controls are poorly designed. Figure 6.8 illustrates how the controls on a lathe are difficult to operate unless the operator has the shape of a gorilla.

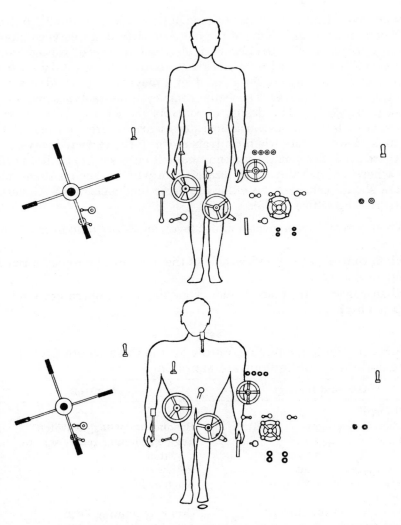

Figure 6.8 The controls of a lathe in current use are not within easy reach of the average man. They are placed so that the ideal operator should be 1372 mm (4½ ft) tall, 610 mm across the shoulders (2 ft), and have a 2348 mm (8 ft) arm span. (SOURCE: This figure appeared in *Applied Ergonomics;* Guildford, Surrey, United Kingdom: IPC Science and Technology Press, Ltd., 27, 1969.) Vol. 1.

There are many different kinds of controls. The choice of a particular one depends on how much force must be used to move the control and how many alternative positions are required (see Table 6.8; Van Cott and Kinkade, 1972).

People usually have expectations regarding how a control should be moved. For example, an electric switch is expected to turn the light on when moved upward. Expectations of this type are called "control movement sterotypes." Some of the more important stereotypes imply that any increase of controlled variable (such as speed, voltage, and rpm) should be initiated by a clockwise rotation for a rotary switch, or a movement downward for a pedal, or a movement upward or away from the body for a lever. These and some other stereotypes are illustrated in Table 6.9. The design of controls should use these preprogrammed inclinations because operators learn to use controls faster and make fewer errors when controls move in the expected ways.

If there are space constraints that make it necessary to position controls where an operator must locate them without seeing them, designers should consider the following error tendencies (U.S. Government Printing Office, 1975):

1. When controls are above shoulder level, operators tend to reach too low.

2. When controls are on either side of the operator, he tends to reach too far to the rear.

3. When controls are placed below shoulder level, operators tend to reach too high.

TABLE 6.8 Recommended Controls for the Case Where Both Force and Range of Settings are Important

Forces and settings	Control
Small force	
2 discrete settings	Pushbutton or toggle switch
3 discrete settings	Toggle switch or rotary selector switch
4–24 discrete settings	Rotary selector switch
Small range of continuous settings	Knob or lever
Large range of continuous settings	Crank or rotating knob
Large force	
2 discrete settings	Detent lever, large hand push button, or foot push button
3–24 discrete settings	Detent lever, rotary selector switch
Small range of continuous settings	Handwheel, rotary pedal, or lever
Large range of continuous settings	Large crank, handwheel

TABLE 6.9 Control Movement Stereotypes[a]

Function	Direction of movement
On	Up, right, forward, clockwise, pull (push–pull-type switch)
Off	Down, left, rearward, counterclockwise, push
Right	Clockwise, right
Left	Counterclockwise, left
Raise	Up, back
Lower	Down, forward
Retract	Up, rearward, pull
Extend	Down, forward, push
Increase	Forward, up, right, clockwise
Decrease	Rearward, down, left, counterclockwise
Open valve	Counterclockwise
Close valve	Clockwise

[a] U.S. Government Printing Office (1975).

It is important to keep these error tendencies in mind when designing control lay-outs so that the operator does not activate other critical controls by mistake.

Once the needed controls have been determined, it is important to code them so that they can easily be distinguished and are easy to operate without errors. Coding implies adding distinguishing features to the controls in one way or another. The designer may code by location, color, size, shape, labeling, or mode of operation (or combinations of two or more of these features).

Coding by location is important. People seem to be very good at remembering where things are located in the space around them (for instance, it is easy to find a light switch in a familiar room, even in the dark). Coding by location provides the opportunity to arrange the controls in functional groups so that controls located together are identical in function, used together in specific tasks, and/or related to one equipment or system component, for example, engine controls separated from boom controls.

When the operator uses several controls in sequence with the same hand, the controls should be arranged in horizontal rows from left to right, in order of operation. If horizontal rows are impractical, the controls could be arranged in vertical rows from top to bottom.

The most important controls should be positioned where they are easily reached. The following factors determine importance:

1. Frequency and duration of use.

2. Accuracy and speed of operation required.

3. Ease of manipulation in terms of force, precision, and speed.

The most important controls should be located at a height between the operator's waist and shoulder and within a radius of 16 in. from the normal working position. They should be grouped together, preferably to the right front of the operator (to be operated by the right hand). When this is not possible, they can also be positioned to the left front. Figure 6.9 shows the preferred locations for controls for a construction machine operator (Society of Automotive Engineers, 1980).

Controls that are used less frequently and are less important should be located within 20 in. of the normal work position. Controls that are used infrequently may be placed to the side. These might be covered, mounted behind hinged doors, or recessed into the instrument panel to reduce distraction and prevent inadvertent operation.

Coding by color is an obvious way of making controls easy to distinguish. Color coding requires that the operator look at the controls. This may increase operation time. Therefore, color should not be used as the only coding technique. The number of colors that can be used are usually limited to the ones that are easy to name and refer to. The most commonly used are red, orange, yellow, green, and blue. Color coding is inadequate in dimly lit environments and color-coded equipment or controls cannot be used in underground or poorly illuminated areas.

Coding by size can make it easy to identify a control both by sight and by touch. Controls that are frequently used should be made larger than other controls. Size coding is frequently used for knobs on the ends of levers. Safe recognition of the sizes requires that the number of sizes used be limited to three. This is especially important for emergency situations, when people are more error prone.

Coding by shape greatly enhances the discriminability of the controls by touch alone. Ideally, the shape of the control should suggest the controlled function. Some examples are given in Figure 6.10. Recognition by touch takes a fairly long time, so shape coding should not be primarily relied upon.

Labeling is an effective way of identifying controls, provided there is adequate illumination to read the labels. There are a few general recommendations for the design of labels (Moore, 1976; Van Cott and Kinkade, 1972):

1. Black letters on a white background are preferred unless the illumination level is below 1 fc or the eye is dark adapted, in which case white letters on a black background are preferred.

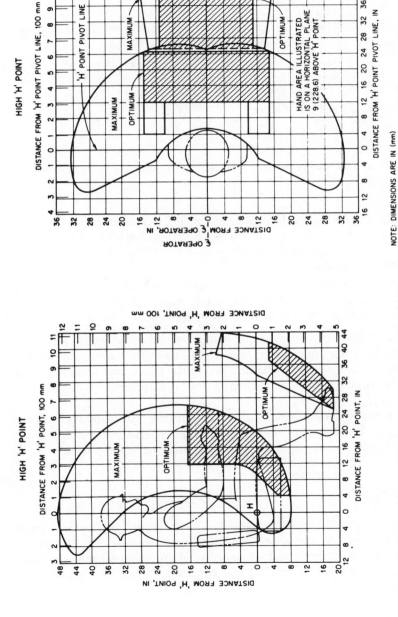

Figure 6.9 Optimum and maximum hand and foot control space for high hip-point (H) location. The figure illustrates a 95th percentile U.S. male construction worker with the seat in the rear position of fore-and-aft adjustment. Provision of 4-in. adjustability accommodates 90% of the operator population. (SOURCE: *SAE J898 Recommended Practice*, SAE, 1980.)

169

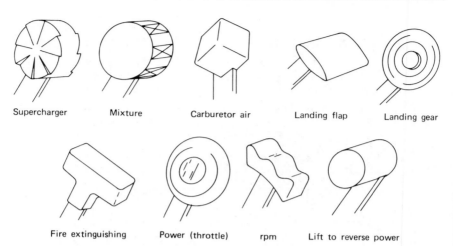

Supercharger   Mixture   Carburetor air   Landing flap   Landing gear

Fire extinguishing   Power (throttle)   rpm   Lift to reverse power

Figure 6.10 Standardized shape-coded knobs for U.S. Air Force aircraft. Some have symbolic association with their functions.

2. Capital letters are preferred with a width to height ratio of 3:5 and a stroke width to height ratio of 1:6 to 1:8. Assuming a viewing distance of 28 ins., the letter height should be at least 0.20 in. Lettering should always be horizontal.

3. The label should be placed on or beside the control. Only common words or abbreviations should be used. Abstract symbols should be avoided since they require special training.

Coding by control resistance or other control movement characteristics can help the operator in identifying controls. Resistance affects the precision and speed of the control, control feel, and smoothness of control movement, thereby reducing the susceptibility to accidental activation.

If the hand is used to manipulate the control, the resistance should be at least 2 lb (this does not apply to finger-operated controls). If the hand and entire arm are used, the minimum resistance should be 10 to 20 lb; for forearm and hand, 5 lb (Dempster, 1955).

6.6.4   DISPLAYS

Displays transmit information about the equipment, job performance, and environment to the operator. Human factors handbooks provide numerous rules concerning how displays should be designed for maximum usefulness (e.g., Van Cott and Kinkade, 1972). This section provides a brief review.

There are three basic categories of display: pictorial, qualitative, and quantitative (see Figure 6.11). A pictorial display gives a symbolic representation of the real world (for example, a display may depict crane position with respect to a structure). A qualitative (check) display is used to indicate the state of a system or situation (for example, a warning light for low oil pressure).A quantitative display presents a number that is a measurement of some aspect of the task or equipment.

There are two kinds of quantitative displays, digital and analogue. A digital display is used if a number is the only information the operator needs. If rate of change is required as well as a current numerical measure, an analogue display is preferred, for example, a moving pointer on a fixed scale.

The purpose of any display is to provide a worker with needed information. Problems of display design should be resolved with human factors in mind.

The first step in the design of a display is to analyze the operator's task and information needs. An operator performing a job has a mental model of what he is trying to do and how he can best achieve his work objectives. Such a model is based on education, training, past experience, and so forth. To design displays that meet the needs of the operator, the designer's efforts must be based on knowledge of the model that the operator uses in his work.

Many of the design guidelines for controls apply to the design of displays as well. For example, displays may be coded by location, color, shape, and label. As with controls, the most important coding principle is location. In general, the locations of important displays must be

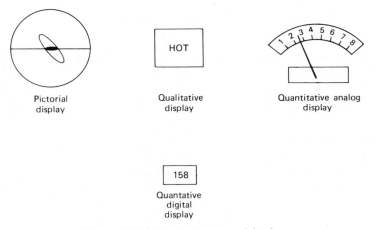

Pictorial
display

Qualitative
display

Quantitative analog
display

Quantative
digital
display

Figure 6.11   Some categories of displays.

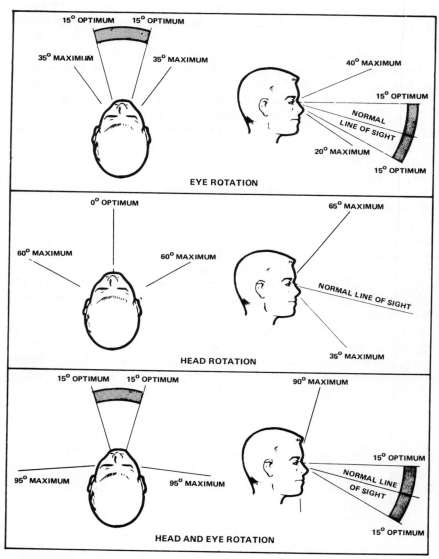

Figure 6.12 Vertical and horizontal visual fields. Observe that the normal (most comfortable) line of sight is depressed 15° below the horizon. The best area for placement of visual information is within a cone ± 15° around the normal line of sight (SOURCE: *MIL-STD-1472B*, U.S. Government Printing Office, 1974).

adapted to the operator's field of view and the characteristics of his eye movements during work. Figure 6.12 illustrates that the normal line of sight is 10 to 15 degrees below the horizon. For optimum display location, displays should be located within a cone that subtends an angle $\pm$ 15° to all sides of the normal line of sight.

The distance from the eye to the display should be approximately 500 to 700 mm (20 to 30 in.). Figure 6.12 also illustrates the visual angles for the limits of the visual field, limits of color discrimination, and limits of symbol and word recognition.

6.7 SELECTION AND TRAINING

Anyone who observes human behavior and performance is struck by the great differences among individuals. This is certainly true of workers' performance. There are obvious differences among workers with regard to the quantity and quality of work done, attitudes, morale, and cooperativeness. Of course, not all differences are unacceptable or undesirable. There is a problem, however, when performance does not meet the minimum standards of the organization. There are three basic ways for ensuring that workers' characteristics and performance will be adequate. These are human engineering, selection, and training. Human engineering (the matching of human capabilities with job requirements) was dealt with earlier. In this section, we review principles for selection and training of employees. Most of these principles apply to management and workers alike.

6.7.1 SELECTION

Today, construction managers seldom use selection procedures other than personal recommendation and probationary employment. When there is a need to establish a reliable work crew, a common practice is to hire and fire workers until the right mix of personalities and skills occurs. Sometimes three or four people are employed before the right person is found for a particular job. This practice creates stressful conditions for both the workers and for management (who must fire the workers that prove to be unsatisfactory). There are several alternative techniques for selecting workers (that do not use this trial and error method). In this section, we review some general principles for selection, the kinds of methods used, and some of the difficulties (practical and theoretical) in using the methods presented. Most of the principles that are discussed apply to the selection of management as well as workers.

Before any selection for a job can take place, the components of the job should be analyzed. This will identify job demands such as specific skills, technical knowledge, and strength requirements. Careful job analysis ensures that the employer knows precisely what is required from the prospective employee. For example, in a small building company there are a variety of jobs to be done by a small number of people. An analysis of the skills needed to do the job will help in defining what kinds of skills and experience are required and what kind of labor can be used. These criteria can then be used to screen applicants.

There are other factors not related to job skill and work experience that might be important to check. Factors such as attitudes about the work, flexibility, and ability to cooperate with fellow workers are all important. Such personality traits might, in the long run, prove to be of greater value to the company than technical skills and know-how.

Selection procedures are often complicated by regulations and restrictions which may make it impossible for an employer to gain access to all potential applicants. Restrictions imposed by the labor unions may reduce the number of applicants; likewise the wages offered and company hiring policies may restrict the possibilities. In the situation where there is only one or a few applicants, selection tests may not be practical or useful.

There are several different methods for screening job applicants. Their common purpose is to try to predict the performance of the applicant at work. The methods used in selection vary greatly among different companies. Some companies rely on interviews alone; others combine interviews with references from previous employers or school records. In some cases, psychological tests are used. Unfortunately, the history of most employers makes it clear that no single selection method is a safe bet; unsuitable employees continue to be hired because accurate predictions about future performance are very difficult to make.

The following methods are frequently used in screening applicants (Drenth, 1976):

1. References from previous employers who are in a position to assess the applicant's past performance and personality.

2. Objective data from the applicant's history such as family background, school records, experience at related jobs, career patterns, geographic origin, and so on.

3. An interview that reveals some of the applicant's personal characteristics and communication skills.

4. A probationary period during which the applicant is observed in the actual work situation or a training situation.

5. Psychological tests of general ability, intelligence, special aptitudes, manipulative skills, etc.

Sometimes a subjective evaluation, such as a job interview, can have good predictive value. This is especially true when the interview is performed by an experienced person. However, research has shown that in a majority of cases, psychological tests are superior to non-scientific, subjective evaluations even when these are made by experienced personnel managers.

During the last 50 years, management psychology has produced numerous written selection tests. There are tests to measure intellectual capabilities, motor abilities, perceptual or artistic factors, interests, values, and attitudes. Some of the tests are very general and applicable in many circumstances (such as general ability and intelligence tests); others are highly specific and designed for particular circumstances (special aptitude tests, work sample tests). The importance of the tests is that scores provide a way to compare applicants' characteristics in an objective, standardized way.

Test scores can be used in several ways. A minimum score may be established for each test; applicants with low scores on any test are rejected. Or the test scores may be combined into a composite score by weighting the individual test scores for their relevance and importance. Likewise, this procedure uses a cutoff score for rejection.

Various kinds of tests can be used to evaluate a candidate during a probationary period. An employer may use criteria such as productivity per hour, quality of a piece of work, absenteeism, tardiness, accidents, and so on.

When there are a number of applicants for a job, the employer faces the question of devising a selection test that will help identify the applicant most likely to perform the job successfully. From the task and job analysis, the employer already knows what is required in terms of physical strength, dexterity, experience, and technical knowledge. These requirements can be used to narrow down the number of applicants. A psychological test may also be used. Finally, an employment interview might be held with four or five of the most qualified applicants. For any of these tests, it is important to prove their validity. Recent court rulings have mandated that any test be directly relevant to the demands of the job.

As was mentioned previously, there is no 100% certain selection test. It can happen that a person with high test scores or a good interview rating is unsuccessful at work. Figure 6.13 illustrates a hypothetical

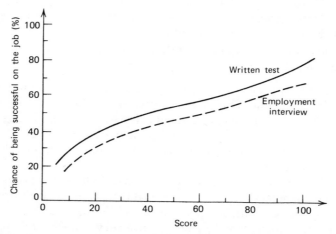

Figure 6.13 Success on the job cannot be fully predicted by any test. Written (psychological) tests are generally better than employment interviews.

example of top test scores leading to success on the job in 80% of the cases. There is, however, also a chance that a person with low test scores will be successful on the job. The figure also illustrates that written tests produce better results than employment interviews.

The imperfection of selection tests is attributable to the difficulty of constructing valid tests, that is, asking the right questions and making certain the questions will be understood and that the answers are easy to interpret correctly. Figure 6.14 illustrates the problems of low validity in selection tests. Imagine that there are two groups of job applicants with 100 persons in each group for a total of 200 applicants. One of the groups is composed of qualified applicants and the other is composed of people who cannot do the work. A perfect test (with a validity coefficient of 1.0) would make the selection of only qualified applicants easy. In reality, however, a validity coefficient of 0.40 is about as good as selection tests obtain. For this hypothetical case, assume a validity coefficient of 0.40. The ellipse in Figure 6.14 defines the boundaries for the test scores obtained. Above the horizontal dashed line is the distribution of test scores for the qualified applicants. Below the dashed line is the distribution of scores for the unqualified applicants.

Depending on the number of applicants the company needs, the ratio of qualified to unqualified employees varies. If 100 applicants must be accepted, there are 63 qualified and 37 unqualified persons selected by the test (Figure 6.14a). But if the company only needs to employ

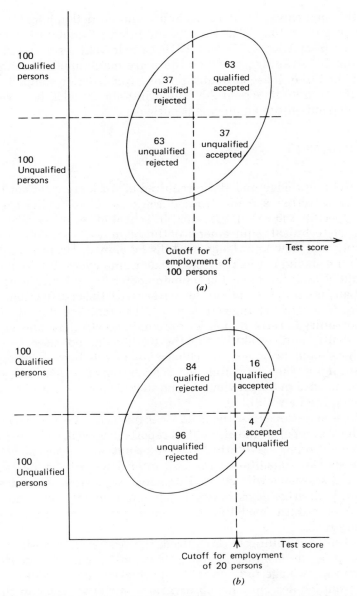

Figure 6.14 Distribution of test scores obtained from a selection test for two groups of applicants, 100 qualified and 100 unqualified. The selection ratio qualified/unqualified applicants improves as the cutoff point is moved from a lower value (*a*) to a higher value (*b*). The number of persons rejected or accepted correspond to a test validity coefficient of 0.4.

20 of the applicants, 16 qualified and 4 unqualified applicants are se-lected (Figure 6.14b). The ratio of qualified to unqualified workers increases from 1.7 (63:37) to 4.0 (16:4). Selection tests are therefore particularly advantageous when there are many applicants and only a few to be hired. It seems doubtful that a construction company would go to the trouble to screen 200 applicants for 20 jobs; however, this might depend on the employment outlook.

6.7.2 TRAINING

All of the knowledge and skills required of work crew cannot be pro-vided by selecting already trained workers. These skills are often highly specific and are not learned in school or on other jobs. In ad-dition, the technical requirements of the job may change substantially, requiring that new knowledge or skills be gained (Drenth, 1976).

Training should not be restricted to new employees. Where possible, it should be an integral part of company policy to make training avail-able at any point in the individual's career with the organization. When major technological changes take place, the employee should be given an opportunity to retrain in order to remain an effective worker. If the job suddenly requires skills that the worker has not used for a long time, he should be given the chance to refresh his knowledge and abilities. If a worker gets promoted to a job that requires abilities that were not needed in his previous job, he should be given the opportunity to develop the new abilities.

Management can assess the need for training by analyzing such data as production reports, quality control reports, feedback from customers, and accident reports (Drenth, 1976). Information can also be collected by independent consultants, through interviews with employees and informal discussions with supervisors, and by questionnaires designed to identify discrepancies between job requirements and individual skills. A procedural model for devising a training program is shown in Figure 6.15.

If a need for training is recognized, the first step in designing the training program is a task analysis. This must accurately identify the skills and knowledge to be taught. Information about the task is ob-tained from job descriptions, observations of work (either in the real work situation or on film), or interviews with workers or supervisors.

Training program objectives are then formulated. This is a fairly subjective procedure and involves determining which skills should be developed and the amount of knowledge trainees should acquire.

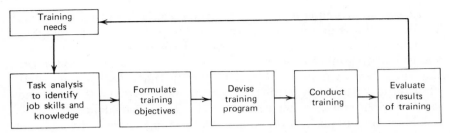

Figure 6.15 Summary diagram of the procedure for devising and evaluating a training program.

The next step is the design of the actual training program. There are several different possibilities and the choice might range from a simple lecture format to the use of highly sophisticated training aids (such as simulator and teaching machines). Each method has advantages and disadvantages, and a considerable amount of information on training techniques is available (e.g., Walker, 1965).

For any training program it is necessary to know at what point the results of training are satisfactory (that is, at what point further training would add costs without returning commensurate benefits). In order to avoid arbitrary decisions on such matters it is necessary to devise tests of trainee proficiency. Each training program should therefore include an evaluation of training results to determine whether or not the training led to the desired effects. This can take the form of proficiency tests given to trainees at the end of the program and will show if they are capable of doing the work for which the training was designed.

As a last step, the entire training program should be evaluated. This will demonstrate whether or not the objectives were adequate and if the training program was effective in meeting the organizational goals.

In order for training to be effective, there are some basic psychological principles that should be considered. If is important that a trainee be motivated to learn and that he or she be given feedback on success in learning. The number and kind of training sessions, and their distribution over time, are also important to the effectiveness of a training program.

Motivation is enhanced if the training material is interesting and if there are positive external factors such as financial or social rewards, praise, and job promotion attached to the training program. Since training is job specific and goal oriented, it is much more motivating than the general learning that occurs in public schools.

The second principle is that the trainee must be kept informed about how he or she is doing and whether or not performance is adequate. Feedback about performance in training increases motivation and makes learning more rapid and effective. The well-known saying that one learns best by mistakes is true only if the individual is told exactly what those mistakes are; this information must be given to the trainee soon after the mistake has occurred. This kind of feedback should be provided continuously throughout the learning process. In addition, the feedback has to be as specific as possible in order to have the maximum benefit for learning. Each mistake must be fully analyzed and the correct procedure explained.

The third important point is that training sessions must be scheduled in an optimal way. If the sessions are too concentrated (too long or too close together), the individual may become physically and psychologically fatigued so that the ability to concentrate and understand decreases while frustration or anxiety increases. There are no general rules for how training sessions should be scheduled; the type of training and the work to be performed dictate what is possible, workable, and efficient.

Any training program must take into consideration that there are large differences in rates of learning and the preferred styles of learning. One of the most fundamental differences is in learning ability. Another major difference is in degree of educational attainment; it is difficult to devise training programs when some employees lack basic education in reading and arithmetic and others are educationally qualified.

Older workers may also have special training needs. Over the past few years evidence has accumulated which shows that the older worker may not be such a bad bet as the common stereotype suggests. Belbin (1965) pointed out that, in performing familiar tasks, age differences are small or do not exist. From this we might conclude that older workers can maintain their performance quite well providing that they are doing their accustomed jobs. Older workers are adversely affected by technological change, however, since their accumulated experience at the old task cannot compensate for the effects of aging on performance of a new task. Any change in method or equipment, however small, should therefore be accompanied by some form of retraining.

Some evidence also suggests that the older worker profits more from a practical, on-the-job type training (the activity method) than from a theoretical (lecture method) type of training. It should be remembered that learning is itself a skill that is maintained only by practice. If employees are given regular opportunities to attend courses and participate in training programs, they will be able to assimilate new ideas more readily and keep up with technological changes with minimum effort and expense (Murrell, 1971).

Much of what we learn in one situation can be transferred to another. This is the fundamental idea on which education is founded—what is learned in school will later be transferred to a relevant context in the real world. The more similar a training situation is to the actual job situation, the more transfer of training can be expected.

Transfer of training can also be negative. When a new skill requires a response that is contrary to what has already been learned in a similar taks, the rate of learning of the new response or behavior is impaired. This impairment is called "negative transfer." The effect of such old habits on new ones can be remarkably persistent. Even though a person has learned the new way of performing the task, he occasionally does it the old way, particularly if he is working under stress (Fitts and Posner, 1967). A practical consequence is that training a worker to use a method or equipment that is similar to those previously used may require more time than expected.

For example, in heavy construction and surface mining, large haulage trucks are used extensively. The drivers usually learn to use a particular kind of truck. The locations of controls on the different trucks may be different. One study found that trucks manufactured by a number of different companies were being used at the same work sites (Conway, et al., 1980). The positions of the retarder and accelerator pedals were reversed on two of the trucks. A driver who is accustomed to one type of truck would therefore experience negative transfer when attempting to drive the other type. Under the stress of an emergency, when the driver must act quickly and by reflex, it is possible that he would use the wrong pedal because it is in the "right" location (the one he is most accustomed to).

REFERENCES

Belbin, M. *Training Methods. Employment of Older Workers 2.* Paris: OECD, 1965.

Chapanis, A. National and Cultural Variables in Ergonomics. *Ergonomics*, **17**, 153–175, 1974.

Conway, E.J., Sanders, M.S., Helander, M., and Krohn, G.S. *Human Factors Problem Identification in Surface Mines*, Vol. I, *Preliminary Findings*. Westlake Village, Calif.: Canyon Research Group, Inc., 1980.

Dempster, W. T. The Anthropometry of Body Action. *Annals of the New York Academy of Science*, **63**, 559, 1955.

Drenth, P. J. D. Selection and Training. In K. F. Kraiss and J. Moraal (Eds.), *Introduction to Human Engineering*. Bonn: Verlag TUV, Rheinland GmbH, Köln, 1976.

Fitts, P. M. (Ed.). *Human Engineering for an Effective Air-Navigation and Traffic-Control System*. Washington D.C.: National Research Council, 1951.

Fitts, P. M. and Posner, M. I. *Human Performance.* Belmont, Calif.: Wadsworth, 1967.

Harris, W. and Mackie, R. R. *A Study of the Relationships among Fatigue, Hours of Service, and Safety of Operations of Truck and Bus Drivers* (Tech. Rep. 1727-2). Goleta, Calif.: Human Factors Research, Inc., 1972.

Helander, M. *The State of the Art of Ergonomics in the Construction Industry* (Tech. Rep. 2707). Goleta, Calif.: Human Factors Research, Inc., 1978.

Hopkinson, R. C. and Collins, J. *The Ergonomics of Lighting.* London: McDonald, 1970.

Kraiss, K. F., and Moraal, J. (Eds.). *Introduction to Human Engineering.* Bonn: Verlag TUV, Rheinland, GmbH, Köln, 1976.

Lindsay, P. H. and Norman, D. A. *An Introduction to Psychology.* New York: Academic Press, 1972.

McCormick, E. A. *Human Factors in Engineering and Design.* New York: McGraw-Hill, 1976.

Miller, G. A. The Magical Number Seven, Plus or Minus Two: Some Limits on our Capacity for Processing Information. *Psychological Review,* **63,** 81–97, 1956.

Moore, T. G. Controls and Tactile Displays. In K. F. Kraiss and J. Moraal (Eds.), *Introduction to Human Engineering.* Bonn: Verlag TUV, Rheinland GmbH, Köln, 1976.

Murrell, K. F. H. Work Organization. *Applied Ergonomics,* 79–91, 1971.

Poulton, E. C. *Environment and Human Efficiency.* Springfield, Ill.: Thomas, 1970.

Roebuck, J. A., Kroemer, K. H. E., and Thomson, W. G. *Engineering Anthropometry Methods.* New York: Wiley, 1975.

Rutenfranz, J., Colguhoun, W. P., Knauth, P., and Ghata, J. N. Biomedical and Psychosocial Aspects of Shift Work. *Scandinavian Journal of Work Environment and Health,* **3,** 165–182, 1977.

Singleton, W. T. *Introduction to Ergonomics.* Geneva: World Health Organization, 1972.

Singleton, W. T. *Man–Machine Systems.* Harmondsworth, Middlesex, England: Penguin, 1974.

Steinbuch, K. "Information Processing in Man." *Proceedings of IRE International Congress on Human Factors in Electronics.* Long Beach, Calif.: 1962.

Society of Automotive Engineers. *SAE Handbook 1980.* Warrendale, Pa.: SAE, 1980.

Tasto, D. L., Colligan, M. J., Skjei, E. W., and Polly, S. J. *Health Consequences of Shift Work* [DHEW (NIOSH) Publ. No. 78-154]. Cincinnati: 1978.

Tasto, D. L. and Colligan, M. J. *Shift Work Practices in the United States.* Washington, D.C.: National Institute for Occupational Safety and Health, 1977.

U.S. Army Missile Command. *Research and Development Directorate.* Redstone Arsenal, Ala.: 1968.

U.S. Government Printing Office. *Military Handbook 759.* Washington, D.C.: 1975.

U.S. Government Printing Office. *Military Standard 1472B.* Washington, D.C.: 1974.

Van Cott, H. P. and Kinkade, R. G. *Human Engineering Guide to Equipment Design.* Washington, D.C.: U.S. Government Printing Office, 1972.

Wilkinson, R. T. Some Factors Influencing the Effect of Environmental Stressors upon Performance. *Psychological Bulletin,* **72,** 260–272, 1969.

Walker, R. W. An Evaluation of Training Methods and their Characteristics. *Human Factors,* 347–354, 1965.

CHAPTER 7

THE PSYCHOLOGY OF JOB SATISFACTION AND WORKER PRODUCTIVITY

SVEN SÖDERBERG, Ph.D.

Department of Industrial Management
Chalmers University of Technology
Göteborg, Sweden

SALENA K. KERR

University of Massachusetts
Amherst, Massachusetts

Two goals of management are to increase productivity and to raise the level of satisfaction workers feel with regard to their jobs. At first one might think these goals would go hand in hand, that increased job satisfaction would automatically lead to increased productivity. In fact, this is not very often the case: Vroom (1964) analyzed 23 studies on this topic and found no systematic relationship between job satisfaction and job performance.

There have been countless and varied attempts to change the working environment in ways that would improve worker efficiency and produce satisfying work conditions. Especially since World War II management has focused a great deal of effort in these areas. A succession of ideas has found acceptance, been implemented in the workplace, and later been discarded as ineffective. In many cases where changes were imposed, basic psychological concepts were overlooked and workers reacted negatively to the changed situation. In other cases, reference was made to psychology but the ideas were oversimplified or misused. It should be possible to design work situations to maximize both productivity and worker satisfaction. Psychology provides some basic ideas that can help management do this.

It is important to begin with a clear idea of *why* people work (there are more reasons than one might think). It is equally important to understand the basis for people's attitudes and feelings about their jobs (satisfaction is only one of many possible feelings). Once we recognize what motivates workers, and understand something about why they think and feel as they do, we are in a position to design or restructure work situations that will be more productive and satisfying.

Management sometimes fails to realize that people's attitudes about their jobs are usually not created at work. Work motivation is actually formed *before* a person enters the job market. The essential motivation to work well is largely created in the social environment outside work; it is influenced by parents, school, friends, the mass media, and so on. Management efforts to boost worker motivation in the workplace may therefore have little effect. Such efforts may even have negative results if the workers recognize a motivation program as psychological "cosmetics." Most people resent transparent attempts to manipulate their attitudes.

Plan of Chapter

Section 7.1 presents basic information on human motivation. The ideas introduced in this section provide the background for a discussion of work motivation. We describe basic human needs (for survival, social interaction, and personal growth). The theories of Maslow, Herzberg, and Vroom are outlined. Maslow and Herzberg were among the first to study nonmonetary incentives to work. Vroom's theory describes how job satisfaction may result from a good match between an individual's needs and the goals of the organization.

In Section 7.2 we present a conceptual model of job satisfaction that combines the theories of Maslow, Herzberg, and Vroom. We describe in detail the way in which conceptual models can help organize information and clarify thinking. From the model that is described in this section we are led to consider the social factors that influence attitudes about work. The effects of social environment are illustrated through a comparison of workers in two different societies.

In Section 7.3 we describe some methods of applying theories in the workplace. It is difficult to go from abstract psychological theories to implementing changes in the work environment. We therefore present some advice for using psychological concepts concerning motivation and attitudes.

Section 7.4 outlines the reasons and rules for group work. We give some directions for organizing group work and for facilitating the needed attitude changes. There are specific attitudes and behaviors that contribute to the success of group work programs; we outline these and provide rules for fostering rapid transition to the group work situation.

7.1 MOTIVATION

Why do people work? Why, in fact, do they do anything at all? The general explanation is that action (behavior) arises from needs that people feel. There are basic, survival needs that motivate us to obtain food, water, sex, and so on. Once physical survival is assured, more complex needs may motivate us to interact with others in groups, achieve social goals, gain higher status, and the like. Friberg (1975) lists four primary motivations for working:

1. *Coercion.*   Historically, people have been forced to work by threat of immediate physical harm. We do not consider this source of motivation further.

2. *Material Rewards.*   Working produces income that provides for basic needs (food, shelter) and, sometimes, luxuries (leisure activities, higher education).

3. *Normative Rewards.*   People may work to gain social or moral approval (as when they work for social, political, or religious causes). These kinds of rewards are strongly linked to a person's culture; for instance, in the People's Republic of China people make collective efforts because of their reverence for the goals of the state. People in other cultures would not experience these kinds of efforts as rewarding.

4. *Intrinsic Rewards.*   An individual may do a job because the content of that job is, in itself, satisfying. These inherent rewards are strongly linked to such factors as personal abilities, skills, interests, and earlier experiences.

It is likely that all these factors can provide motivation for work. The relative importance of each kind of reward varies from person to person and from one set of circumstances to another.

Although practically everyone uses the word "motivation," there is no good, generally agreed-upon definition. Most people have some idea what it means; psychologists argue for one meaning or another, and some doubt the very existence of "motivation" in human beings. For our purposes we think of work motivation as the desire or ambition to work well. In practical situations the level of work motivation has been gauged by productivity, worker health, absenteeism (especially short-term absenteeism), job satisfaction, and attitudes concerning the work, co-workers, and self.

The question of motivation would be much simpler if individuals did not differ in so many ways. Figure 7.1 is a simple diagram of the complicated relationships between an individual's mental processes and characteristics, motivation, and behavior. Objectively, the environment might be exactly the same for two people: say, they work in the same place, doing the same job, for the same boss and the same pay. Why would one be more satisfied or productive than the other? Possible answers are given below in terms of Figure 7.1.

Inherited Characteristics

We all inherit a set of capabilities and limits in the same way we inherit eye color and height from our parents. Though training and practice can improve people's performance at work, each person has a different *maximum* capability in any given area of human effort. This maximum capacity to do any particular thing is an inherited characteristic (for example, an upper limit on physical strength or reaction speed, or the ability to do some kinds of math problems or

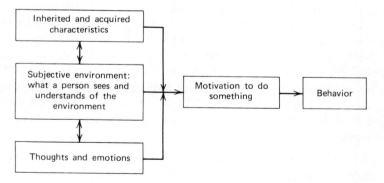

Figure 7.1 Motivation and behavior are affected both by our inherited and acquired characteristics and our logical and emotional evaluation of the environment.

memorize music easily). In the case of two people doing the same job, perhaps one is unable to work as fast as the other and feels frustrated about it. It is essential to keep in mind that inherited capabilities can be maximized by training or suppressed by lack of training and disuse.

Acquired Characteristics

These are the capabilities and limits we get from our social setting and education. Children learn parents' habits of speech and thought. Later, in school, each child learns a different amount according to his or her capacities and the opportunities available. The inherited characteristics of every individual develop in a unique social environment; the specific forces that act on each of us depend on our parents, friends, economic background, educational opportunity, exposure to mass media, and so on. You can see that these complex social forces acting on individuals (who were already very different at birth) result in an extremely varied population. In the case of the two workers discussed above, perhaps the one working more slowly holds the attitude that the job is beneath his or her status. This kind of feeling might be learned from parents or friends, and would create great dissatisfaction with the work.

Subjective Environment

Because each of us has a unique combination of characteristics, we all see the world in different ways. Our physical systems for perceiving the environment are different; for example, some people are near-sighted, and some can taste sweet flavors that others cannot taste. In addition to this, the way we perceive a situation is strongly affected by what we have learned (our attitudes, thoughts, and feelings as well as the facts at hand). What each person sees and understands of the actual environment is called the "subjective environment." For example, the work environment of a worker who feels "above" the job is quite different from the subjective environment of a satisfied co-worker. The subjective environment of the first worker is quite unpleasant, and he or she will be reacting to this subjective perception rather than to the actual situation. The satisfied worker is also reacting to a subjective environment, since no one can perceive a situation objectively all the time. The difference between the two is that the first worker's attitude about the job changes the way the work is perceived in a very negative and undesirable way.

Thoughts and Emotions

These are internal; they are hard to measure and hard for people to express or even identify (at times). But these mental processes are important in determining motivation. A person forms thoughts on the basis of experience and learning; emotions affect those thoughts in a number of familiar ways. Both the thoughts and emotions help determine what a person feels motivated to do. Sometimes thoughts and emotions are conflicting: an individual might think it wise to do one thing but feel strongly like doing something else. The motivational state is a balance between the two; a decision to satisfy the greatest need.

In the case of the two workers, perhaps both would agree that it is important to work well. This is only logical; it is a thought that many people share. But the emotions of the dissatisfied worker may conflict with the idea; he or she may feel resentful toward an insensitive supervisor or angry over being passed over for promotion. These emotions might lessen the motivation to work.

In sum, Figure 7.1 shows that individuals' characteristics determine what they can perceive of the objective environment. The subjective environment (all of a person's perceptions of a situation) is influenced by thoughts and feelings. All of these factors help create the motivation to do something. When the motivation is strong enough, the person acts on it.

7.1.1 BASIC HUMAN NEEDS

Usually, motivation is associated with a felt need. As mentioned before, when there is a conflict between needs, a person generally tries to satisfy the most important one. Table 7.1 is a representative list of needs; each may give rise to any of a variety of behaviors or feelings. The list may not be complete, and experts may argue whether one or another "need" is correctly included in any such list. But it gives us a starting point for identifying specific reasons why people act and feel as they do. This should lead us to a clearer notion of why people work and how they might become happier or more productive in their jobs. Keep in mind that physical survival and comfort have priority when people experience conflicting needs. Once survival and physical well-being are assured, people are generally motivated by social and self-actualization needs.

People experience all or most of the needs listed in Table 7.1 at different times in their lives. Their emotional and behavioral reactions

TABLE 7.1 Basic Human Needs

I. Physical survival and comfort needs
   Breathing
   Hunger
   Thirst
   Sexuality
   Maternal behavior
   Harm avoidance (avoidance of pain, cold, heat, etc.)
   Elimination of waste
   Physical activity
II. Social needs
   Fear (avoidance of threats)
   Aggression (appropriately expressed)
   Affiliation
   Power
   Achievement
III. Self-actualization needs (personal growth)
   Experience
   Curiosity (exploration)
   Creativity

to things are determined by the balance of needs they feel. For example, when a workplace is reorganized, some people react favorably and some react negatively. It is difficult to propose any change that will meet with everyone's approval. When people react differently to the same situation it is because their needs, their motives at that time, are different. This is a good reason to study motivation in sufficient detail so that we can predict people's reactions, organize work to meet many of their needs, and avoid negative responses.

Physical Survival and Comfort Needs

These are almost self-evident. People have clear needs for food, water, air, sex, mothering, and the elimination of wastes. Somewhat less evident but equally important are needs to avoid physical damage and to get sufficient exercise. All these needs are primary and contribute to individual and species survival.

Social Needs

Fear serves a protective function in threatening situations; these used to involve physical survival more but now social threats are the ones we fear more often. Aggression may once have contributed to physical survival as well; now it can produce states of conflict because people must restrain aggressive impulses for the good of society.

The need for affiliation is a need to join other people in groups that are bonded in some way. A person with a strong affiliation need searches for opportunities to exchange thoughts and experiences with other people and to participate in group activities.

A need for power is the urge to control social situations. People may express this need by requiring decision-making authority or freedom from imposed schedules at work. A strong need for power might lead someone to choose a career in politics, journalism, teaching, and so on.

An achievement need leads people to seek situations where they can work toward an established goal. A feeling of achievement is important to most people; it is the feeling that they have worked well, accomplished something worthwhile and received appropriate recognition for it.

Self-Actualization (Personal Growth) Needs

The term "self-actualization" is taken from Maslow (1954) to describe the drive to fulfill our potentials, to develop and use our personal capacities, and to experience growth. When the physical and social needs are satisfied, a person may turn inward and recognize a need for further development. This may express itself in a search for new experiences, curiosity about previously unexplored topics, or efforts to create, artistically or in other ways.

All this information about basic human needs does not directly answer the question of how to make work more satisfying or meaningful. Our main objective here is to point out that people have varied needs that make them more or less satisfied with a particular work situation. These differences make them react differently to the workplace. In the next section we outline somed theories of work motivation that illustrate how basic needs may affect work. The ideas of Maslow, Herzberg, and Vroom illustrate the development of concepts that may be applicable in the workplace; these concepts are fundamental to understanding more recent research and thinking on the motivation to work.

7.1.2   MASLOW AND HERZBERG: EARLY THEORIES OF WORK MOTIVATION

Maslow and Herzberg were among the first to make clear the importance of nonmonetary incentives for increasing work motivation. The theories of both men are based on the idea that there are basic human needs that must be satisfied and that many of these can be satisfied in the workplace. The emphasis of both theories is on social and self-actualization needs, since these may be the most relevant to today's working situation.

Maslow (1954) divided basic needs into five categories: (1) physiological needs, (2) security needs, (3) needs for belonging and social activity, (4) esteem and status needs, and (5) need for self-actualization. (You can see that this roughly parallels the longer list in Table 7.1.) Maslow described human needs in terms of a needs hierarchy; an individual proceeds through life by satisfying first the lower (physiological) needs, then advancing, a step at a time, through the satisfaction of higher needs. According to Maslow, the individual cannot ascend to a higher step without having satisfied the needs at lower levels (see Figure 7.2).

The consequences of Maslow's theory are that, once the worker has sufficient financial security to sustain life, other kinds of incentives

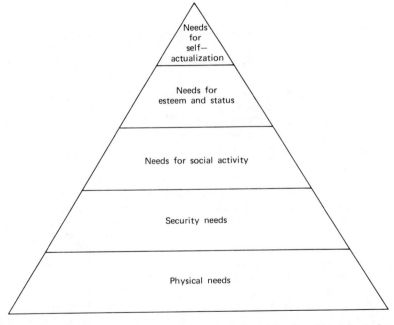

Figure 7.2 Illustration of Maslow's model for need satisfaction. Physiological needs and security needs are the most basic. When these have been satisfied, the individual can focus on satisfying the higher order needs.

must be available if he or she is to continue to be motivated at work. The need for belonging and social activity may be answered in the workplace by creating group work conditions (a later section deals with this more fully). In the future, management may have to consider still other needs (for esteem and self-actualization) in the design of jobs and working conditions. Much has already been done to enhance particular jobs by enlarging responsibility and increasing variety for workers in various occupations.

Although Maslow's theory emphasized an individual's development, Herzberg, et al., (1959) was primarily concerned with attitudes toward work. He conducted surveys to determine how large numbers of workers felt about aspects of their jobs such as opportunities to take responsibility, job security, and pay. From the answers to his questionnaires Herzberg identified two kinds of factors that affect job satisfaction; he called these "motivation" factors and "hygiene" factors. Table 7.2 lists the factors Herzberg determined to be important. The motivation factors determine how well a person likes a job, whereas the hygiene factors concern the more concrete aspects of working life (such as physical working conditions, pay, and so on).

TABLE 7.2   The Theories of Herzberg and Maslow

Herzberg	Maslow
Motivation Factors	
The work itself	Self-actualization
Opportunities for achievement	needs
Opportunities for growth	
Opportunities to take responsibility	
Advancement opportunities	Esteem and status
Recognition	needs
Status	
Hygiene Factors	
Interpersonal relations with supervisors, co-workers, and subordinates	Social activity and social involvement needs
Technical supervision and support	
Company policy and administration	Security needs
Job security	
Working conditions	Physical needs
Salary	
Effects of work on personal life	

Table 7.2 illustrates the theories of Maslow and Herzberg together. Herzberg's motivation factors correspond to Maslow's concepts of the need for esteem and self-actualization. A workplace that is substandard in physical terms (for instance, one with high levels of noise or air pollution) might be quite acceptable to workers if the work offers motivation factors such as a feeling of achievement or the opportunity to take on new responsibilities. These benefits act as incentives because they answer the needs for self-esteem and personal growth. Workers can overlook a lot of negative aspects of a work situation if the right kind of positive factors is there.

Herzberg's hygiene factors correspond to Maslow's concept of "lower" needs. These are called "hygiene" factors because they contribute to mental and physical hygiene or well-being. A company that provides decent physical working conditions, security, reasonable pay, good management, and opportunities for social expression is taking care of these factors. But, according to the theory, these alone cannot increase job satisfaction. They can only prevent dissatisfaction.

The parallels between the two theories might lead one to think that it is a simple matter to fulfill the needs identified by Maslow by concentrating on the specific factors indentified by Herzberg. In practice, this can rarely be done. Like all theories, Maslow's and Herzberg's have value in organizing our thoughts on a complex problem. Still, there are too many other variables in the multitude of practical situations to apply these simple concepts with reliable success (although it might well be worth a try).

The theories of Herzberg and Maslow help explain why people behave differently in companies that manage motivation and hygiene factors differently. Gardell (1971) called the motivation factors "stay" factors and the hygiene factors "quit" factors. In other words, work motivation programs can fail even when companies spend considerable effort and money creating good physical working conditions if they do not pay attention to the motivation factors. An example is the Volvo Company, which spent millions to improve hygiene factors but did not succeed in increasing motivation or productivity.

On the other hand, in some industries where physical work conditions can be downright unpleasant, workers may be highly motivated. The construction industry provides an example: because of the nature of the work it is difficult to get rid of excessive noise, air pollution, and so on. But the work environment may be rich in motivation factors; each worker makes a tangible contribution to the building and is recognized for performing skilled and necessary work. The work is done by small groups that communicate freely and have a good deal of decision-making autonomy. Needs for affiliation, esteem, responsibility, achievement, and recognition are fulfilled. This is probably more important to many individuals than the need for quiet and healthful working conditions.

7.1.3 RESEARCH AT THE TAVISTOCK INSTITUTE

Work at the Tavistock Institute focused on still another set of factors that influence work motivation and job satisfaction. Job satisfaction and even performance may be limited by the design of the workplace. A case in point was outlined in a study by Trist and Bamforth (1951). They examined the results of reorganization and mechanization of work in British coal mines. New technologies were adopted that made it necessary for miners to work in groups of 40 or 50. Before the reorganization they worked in small, independent groups. Each new, large group was assigned to one of three separate areas of responsibility (preparation for mining, or transportation of coal). The jobs were separate in time because they had to be done in sequence, so no interaction could take place between workers doing different jobs. Each large group was expected to do its assigned work, then leave, to be followed by the next large group doing the next job in sequence.

The reorganization of work was intended to increase efficiency; management designed the new structure to make use of new technologies. However, the new structure produced conditions of limited communication; this was not foreseen by the planners who implemented the changes. The planners thought the new system would be more efficient than the old one because new technologies are frequently assumed to be better than existing ones. Klein (1974) summarized the results of reorganizing the workers to fit a new technological environment:

> Each of these teams or workers, optimizing conditions for itself, created and passed on bad conditions to the work groups responsible for the subsequent tasks. Instead of enabling them to cooperate with one another, the new system created irresolvable conditions for interpersonal and intergroup conflict. resulting in competitive individualism, mutual scapegoating and a high level of absenteeism, all of which contributed to a low level of productivity. At the same time, all controlling and coordinating activities had to come from outside and above the teams, since no one at the workplace knew the whole story.

The point of the example, which is also the emphasis of the Tavistock investigations, is that the technical system of the workplace interacts with the social system. If the technology is merely mechanically efficient it may actually destroy the conditions necessary for communication, cooperation, and a high level of worker motivation.

Based on the findings of the Tavistock Institute, the Norwegian investigator Gustavsen (1969) made some recommendations concerning work design. These are listed in Table 7.3; each recommendation aims at designing work environments that promote (rather than block) job satisfaction by fulfilling workers' needs.

7.1.4 VROOM'S MODEL OF WORK MOTIVATION

Vroom's investigations led him to a slightly different concept of work motivation than those already described. The model he developed suggests how an individual may become motivated by expectations about a job, and how the worker's motivation may change as expectations are met or disappointed at work. Vroom (1964) describes a continuous matching process in the mind of the worker: the individual compares his or her own needs with the organization's goals or needs. From this comparison workers more or less estimate the likelihood that their needs will be satisfied at work. Figure 7.3 diagrams this matching

TABLE 7.3   Work Design for Increasing Job Satisfaction

1. Each job should be as varied as possible (including more than one task).
2. A job should consist of a pattern of activities meaningful to the worker.
3. The optimum work cycle length should be established (this is the length of time to complete the sequence of tasks in the job).
4. The worker should help to establish the criteria for quantity and quality of production.
5. The worker should help to establish the criteria for evaluating his or her job performance.
6. There should be opportunities to include voluntary, additional tasks in the job (for instance, maintenance or administrative work).
7. The worker should have a clear idea of the importance of his or her job to the end result or product.
8. The worker should have opportunities to do tasks that are respected and have status.

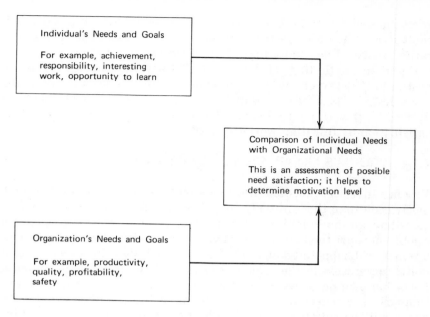

Figure 7.3   Vroom's model of motivation.

process. The end result is a predicted level of job satisfaction; if the level of expected need satisfaction is high, the worker will be motivated to work well.

Vroom's model emphasizes the importance to the individual of knowing well what his or her needs are. It is often the case that people have only a rather fuzzy picture of what they really need and want. The model also demonstrates the importance to management of clarifying and stating the organization's needs to the worker (especially prior to employment). When the goals of the worker are a good match with those of the company, a high level of motivation can result.

7.1.5 SUMMARY

1. There is no clear relationship between job satisfaction and productivity.
2. Job satisfaction is determined by the following:
   Need satisfaction on the job.
   Expectations about need satisfaction and the work situation.
   Need strength.
   Characteristics of the work situation.

The individual's experiences.

Social factors that determine attitudes about work.

Inherited and acquired characteristics of individual workers.

3. There are three groups of basic needs: physical needs, social needs, and the need for self-actualization.

4. Maslow viewed motivation as a hierarchy of needs with physical needs at the lowest level and self-actualization at the highest.

5. Herzberg viewed work motivation as created by motivation (stay) factors and hygiene (quit) factors.

6. Findings of the Tavistock Institute show that the technological environment can interfere with efficient organization of the social environment at work.

7. Gustavsen described eight ways to design work to increase motivation.

8. Vroom viewed motivation as an ongoing process of matching the individual worker's needs with the company's requirements. The individual estimates the probability of satisfying his or her own needs by working at a particular job.

7.2. CONCEPTUAL MODEL FOR JOB SATISFACTION

Many efforts have been made in various industries to increase job satisfaction and to provide conditions for more efficient production. A few of these efforts have already been mentioned. Some have been successful and some have not; many were implemented in a "shotgun" attempt to solve a problem. That is, the ideas and programs were developed without any guiding framework of ideas for understanding what conditions might create job satisfaction. Even when such a program succeeds, it is difficult to say exactly *why* it succeeded or to use that information in a different situation.

A conceptual model (such as those diagrammed in Figures 7.1 to 7.3) provides the framework of ideas we need for analyzing complicated problems. A model may help us predict how well a proposed solution will work in a given situation. Industry has experienced two major problems in trying to implement changes *without* a guiding theory or conceptual model: (1) Management ideas about the nature of the problems were rarely clear, making it difficult to pinpoint solutions. (2) No follow-up programs were designed to check on the effectiveness of the changes in solving the problems.

A conceptual model is basically a tool to clarify thinking. Models such as the ones we have already discussed are the simplified results of much research and thought on complex issues. It is convenient to outline these psychological concepts in simple flow diagrams that make the main points easy to identify and remember. One important point becomes clear from looking at all the models researchers have created: there is no single, absolutely correct way to look at human behavior. People are too varied and situations too diverse for any simple model to predict exactly how people will react to anything, or when they will be motivated and when they will not.

The model for job satisfaction outlined in Figure 7.4 results from combining several theories and concepts (some of them already familiar). In this model, job satisfaction (at the bottom of the diagram) is thought to be determined by the interactions of all the other factors shown. Some of the ideas of Maslow and Herzberg can be recognized in boxes A, B, E, and G. Vroom's concepts enter the model in boxes D, F, and G. Another major variable determining job satisfaction is in box C; the effects of social factors are so important that we consider them at length in the next section.

What follows here is a brief description of each of the factors named in Figure 7.4. In reading and thinking about the model, one should

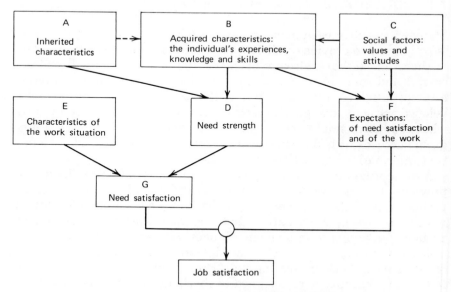

Figure 7.4 Determinants of job satisfaction.

remember that a change in any of the determining factors (boxes A through G) will result in a change in the level of job satisfaction. The determining variables may be thought of as feeding into the outcome; each factor affects job satisfaction to a greater or lesser extent depending on the individual and the circumstances.

Job satisfaction is a gauge of attitudes and feelings about work. These are usually measured byd attitude surveys. Positive attitudes indicate job satisfaction. A positive attitude is reflected in a statement such as, "I enjoy working here, the people are nice and the company is fair."

Box A: *inherited characteristics* have already been discussed. These are biologically determined capabilities and limitations. We inherit a basic physical self with potential to develop to a certain level. Inherited characteristics also partly determine the kinds and relative strengths of a person's needs, although it is difficult to separate which needs are inborn and which are learned later in life.

Box B: the individual's experiences, knowledge, and skills are *aquired characteristics*. These are the sum of a person's interactions with people and situations throughout his or her life. Individuals learn from every experience, from earliest childhood on, about what attitudes and actions produce the best results. All these experiences produce the background for each person's specific (and different) profile of needs, goals, and expectations.

Box C: *social factors*, that is, values and attitudes, are communicated to people in many ways from family, school, friends, television, and the like. The people we associate with most frequently are called a "peer group." Peer groups constantly exert pressure on members to act and think in certain ways. Friends and respected co-workers have a great deal of power in influencing an individual's attitudes, including those toward work.

Box D: *characteristics of the work situation* include every aspect of the work environment. These are the ecological conditions of the workplace, the whole system of which the worker is one part.

Box E: *need strength* was discussed in an earlier section. The model shows that the strength of any need is determined by inherited and acquired characteristics (we learn what we need and how important each need is from early experiences).

Box F: *expectations* of need satisfaction and expectations about the work situation are mainly determined by what an individual has learned to expect from other experiences in life. The beliefs, attitudes, and values of society and of the worker's peer group also affect his expectations. To some extent, need strength may bias a person's expectations (for instance, "wishful thinking" may enter into the situation if a person needs something very much).

Box G: *need satisfaction* through work may be difficult for the individual to recognize or analyze precisely. A worker is simply pleased with a job in general, without being able to name exactly which needs are being satisfied. If need satisfaction is adequate, a worker feels that the job is a good one. Where need satisfaction is insufficient, a worker may express resignation or feelings of alienation from the work.

7.2.1   SOCIAL FACTORS THAT AFFECT JOB SATISFACTION

In this section we discuss more fully the factors in boxes B and C of Figure 7.4. Up until now we have been looking at job satisfaction in terms of individual characteristics, needs, and expectations. All these are important and all are strongly influenced by the society in which a person lives and works.

Job satisfaction may be thought of as part of an overall attitude about work. Rokeach (1965) defines an attitude as follows: "An attitude is a relatively enduring organization of beliefs about an object or situation predisposing one to respond in some preferential manner." An attitude about work, then, involves everything a person has learned and believes about working; these beliefs lead a person to respond positively or negatively to the work situation. We would expect work motivation and performance to be affected by the worker's attitude; a worker with a negative attitude toward work many never experience satisfaction at a job no matter what the task or conditions. From this we conclude that to increase job satisfaction it may be necessary for attitudes to change.

To understand how attitudes about work might change, it is useful to think about where they come from in the first place. This is where the factor of the social environment comes in: a person's attitudes generally come from other people, from the society in which the individual lives. Attitudes are influenced by (or even adopted directly from) family, friends, what is learned in school, and every other significant contact with people. These contacts are not always direct. Mass media such as television, newspapers, magazines, and the like represent indirect contacts with others that can be very powerful influences on the individual's attitudes.

Within a society many different attitudes toward work may exist. For instance, the protestant work ethic has been dominant in the United States for many years. This attitude toward work emphasizes the value of hard work, dedication to the goals of the organization, and providing financial security for the family. In the last few generations, people have adopted other attitudes about work: the "hippie" work ethic is an example (to contribute as little as possible to the established economic system). Most people in the United States today have an attitude toward work that is somewhere between the two extremes: they are not willing to put so much into a job as before without greater rewards. (Neither are they willing to join the counterculture and live without the benefits of a steady income.) By contrast, in developing countries social conditions make it necessary for people to work long and hard for relatively low pay. People in these societies have rather different attitudes toward work, since they are working for physical survival rather than increased spending power for consumer goods. Those that are employed may not like their jobs, but they often work as well as possible because they cannot afford to lose them.

A particular worker's attitude depends on what he or she has been exposed to (life experiences). The general attitudes in a society are filtered down to each individual through the network of people in his or her local social environment. A set of values is learned from the family; the nature of these values (including an attitude toward work) depend on where the family lives, how well off they are, their ethnic background, and so on. In the same way, an individual is likely to learn attitudes and values similar to those of friends, schoolmates, and. co-workers. Since each individual has a unique life history, each comes to the job market with a slightly different set of values and a definite attitude about working.

To illustrate the effects of social factors on job satisfaction, we compare workers from two different countries. Sweden and Korea contrast sharply in social values and in attitudes about working. Sweden is a highly developed country with a high standard of living; the social philosophy there is humanistic. With respect to work, most Swedish people feel that one should not work too hard or too long, that leisure is more important than work, that physical working conditions should be healthy, and that a job should provide opportunities for personal achievement and growth. Korea, on the other hand, is a developing country with a comparatively low standard of living. The social philosophy in Korea emphasizes the need for material growth and the need for every individual to work hard to increase the general level of material well-being. Workers are publicly honored for outstanding sacrifices and hard work. Great values is placed on making a contribution to society rather than on achieving personal goals such as self-actualization.

A consequence of the Swedish values and attitudes is that it is difficult for a worker to experience satisfaction from doing a job; most people work only to produce income and find no pleasure in the work itself. Sven's case illustrates the problem: he is a sheet-metal worker in his middle twenties. Although his work is occasionally challenging and demands considerable craftsmanship, he does not think it is worthwhile and feels no sense of achievement in his job. Discussions with his friends and coworkers focus on hazards such as getting poisoned by welding smoke or falling from a roof. There is never a discussion of the positive aspects of the work. In addition, Sven started his working life with very high and unrealistic expectations of the financial and personal rewards a job would bring. He is constantly disappointed in the "real life" situation because the expectations he learned from family, school, the mass media, and such were much too high. Because he reacts negatively to the job, Sven's boss thinks he is a "spoiled kid with no sense of responsibility." This, of course, simply increases Sven's dissatisfaction.

The Korean sheet-metal worker, Kim, is in complete contrast to Sven. Whereas Sven works moderate hours for pay that supports a comfortable lifestyle, Kim works long and hard for very low wages. He is often physically exhausted when he gets home because he usually works 2 to 4 hr overtime every day. Yet he likes his job and has a strong sense of achievement about his work. Because of the attitudes he has learned, Kim feels he has contributed greatly to the benefit of his family, his country, and himself. The more and harder he works, the greater the self-respect and accomplishment he feels. In Sweden, this would run contrary to accepted ideas: a man would be thought very stupid to work like that. By the same token, in Korea Sven would be regarded as disloyal and ungrateful; he would probably not be able to keep a job because his attitudes would be so incompatible with the situation.

These examples show how the perception of a job depends on the attitudes workers learn from their social environments. The effect on job satisfaction is clear. Now the question remains as to how attitudes might change to allow workers like Sven to feel some involvement with their work and some satisfaction in performing it well.

7.2.2 ATTITUDE CHANGE

In spite of the considerable power of the social forces that form a person's attitudes, it is possible for negative attitudes about work to change. It is especially useful for management to know how to encourage and facilitate beneficial changes in attitude, particularly when there is to be a change in the work situation. When a workplace is reorganized, worker attitudes are often negative; people resist change and prefer what is familiar to what is not. Even when people know in their minds that a new system may be better, their attitudes and feelings may be negative simply because of lack of familiarity.

This points up once again the fact that people do not always see things rationally and objectively. Workers perceive the work situation subjectively, in a way that is biased by personal attitudes and needs. In the case of a new job or the reorganization of work, people rely primarily on associates and co-workers in forming their reactions. Management must consider this in trying to encourage positive reactions; a beneficial attitude change must originate on the shop floor, among co-workers. A new attitude can never simply be handed down from above. A truthful, intelligent education campaign to sell workers on a new idea may help ensure their acceptance of it. Likewise, support of supervisors and union people for a new system may generate support among the workers who will use it. Yet the workers' immediate social environment, their peer groups, will be the main determining factor in deciding the overall response to something new.

Attitude change also accompanies changes in habits and behavior. When a person (voluntarily or not) changes from an accustomed way of doing things to an unfamiliar way, attitudes begin to change. The new way eventually becomes familiar, habits are changed, and, after this happens, the person's attitude becomes more positive. In the case of work reorganization, initially negative and counterproductive responses may slowly change into enthusiasm if the situation is, handled well by management.

People want their attitudes to be consistent with their actions. So once they start doing something, they begin to approve of it more and more. An example is the use of seat belts in cars: in recent years, the use of seatbelts has been made mandatory in several countries. At first there was great resistance and some people were completely negative about changing their behavior (no matter how beneficial the results might be). As time went by and people got into the habit of using seat belts, though, they became increasingly positive about using them. That is, their attitudes became consistent with their actions. Another example is a worker in an unskilled occupation who, despite having a low status job, enjoys the work and feels a sense of accomplishment. The basic reasoning of such a worker is, "I am not an inferior person, so this must not be an inferior sort of job. I wouldn't be doing that kind of work, because I am not that kind of person."

It would be difficult to overstate the importance of group membership on attitudes and behavior. For example, a worker who is a member of a work group develops a sense of identification with the group. This identification means that the worker adopts the ideas of the group and feels motivated to work for the group's goals. The group might be a small number of co-workers that the individual sees every day. But there is no reason that a worker's identification should stop there. If management policy is wise, individual workers also identify with the company and adopt company goals. Positive attitudes toward the larger "group" creates motivation to work. If managers understand how workers perceive their jobs and the work environment, they can create the sort of environment that fosters identification with the company.

Supervisors and other opinion leaders (union people or highly regarded workers) play an important part in helping the individual understand his or her role as a member of the organization. If supervisors and work group leaders are sensitive to workers, they can promote beneficial perceptions of the work, develop agreement on what attitudes and needs are appropriate, and define how workers should relate to the work environment. Supervisors and opinion leaders can also make the company's goals more real and motivating for individuals who feel no sense of identification with the distant front office. They can create an environment in which company values and expectations are always at least somewhat in mind—not resented, but accepted as reasonable and fair.

Sociologists have a name for a person who can encourage others to accept change; he or she is called a "change agent." If an opinion leader is identified among a group of workers, that person can become a valuable assistant in implementing change. When workers hear about new ideas from one of their own, they are less threatened by the lack of familiarity. Also, the opinion leader can express ideas in terms the other workers understand, the greatly facilitate the program through knowledge of how things really work on the shop floor. It is essential that the change agent be thoroughly informed of the change and genuinely convinced of its merits.

These general ideas are not new. Management efforts to get employees to identify with the company and adopt positive attitudes are numerous (by means of selection, training, uniforms, titles, company newsletters, and the like). These efforts may pay off as long as the workers regard the organization as trustworthy. If this trust is undermined, communication from management is regarded with suspicion and often has an effect opposite that intended.

7.2.3   SUMMARY

1. A conceptual model clarifies thinking and organizes knowledge concerning complex problems.

2. One conceptual model shows how job satisfaction is determined by the following:

Inherited characteristics.

Acquired characteristics (experience, knowledge, and skills).

Social factors.

Characteristics of the work situation.

Need strength.

Expectations (of need satisfaction and about the work).

Degree of need satisfaction by the work.

3. Social factors affect job satisfaction because they influence people's attitudes about work.

4. Attitudes about work can be changed; there are various ways that new attitudes can be encouraged.

7.3 APPLYING THEORIES IN THE WORKPLACE

So far we have been discussing rather abstract psychological theories and models. It is clear that most of these have not been tested in the wide variety of work situations that exist in the world.Without such testing the theories and models cannot be said to predict reliably what will happen in each specific, practical application. So how can they be used to design or reorganize work?

First, even a very general model can be useful in organizing an approach to work design; it can help structure all the complicated psychological factors that must be considered. To be used this way, two requirements must be met:

1. The model must be sound. That means it must be relevant to the situation, fairly general in scope, consistent, and easy to communicate. Sometimes a simple but reasonable model with some theoretical flaws is better in practical applications than an elaborate, complex model that only a few people understand.

2. The practitioner must be skilled in using the theory and able to invent specific ways to apply it in a particular work situation. The practitioner must be objective in assessing the results of these efforts and flexible in thinking about both the model and the situation. In some cases it becomes necessary to admit mistakes and analyze what went wrong; a very open mind is required for this.

An example of the simplified kind of model we are describing is Maccoby's (1978) model of job satisfaction. Maccoby claims that satisfaction at work depends on four interconnected factors; his model is illustrated in Figure 7.5. In Maccoby's terms, "security" means both job security and confidence in the safety of the physical work environment. "Participation" means that the worker influences planning and has a role in decision making. "Equity" means the worker feels that supervisors are fair, that no one plays favorites or singles out a scapegoat when things go wrong. "Self-actualization" is the same as Maslow's concept—the work situation provides opportunities for the worker to develop and grow.

No model such as this is "true" in the scientific sense. Maccoby's model is greatly oversimplified; he uses concepts from a number of theories without really explaining them or claiming that the model is complete. Nonetheless the model is useful: it could serve as a starting point for designing work or as the basis for negotiations between workers and employers. For each practical case, what is actually meant by "security," "participation," and so on would be differently defined by the people involved. For instance, worker involvement in decision making is different in a construction crew than on an assembly line, simply because the work itself and the decisions to be made are vastly different. But the principle of the model, that workers should have a role in decision making, remains valid.

7.3.1. LEARNING TO APPLY THEORIES

Applying psychological theories in the workplace is a demanding and difficult job. Nothing about it is clear-cut; everything the practitioner

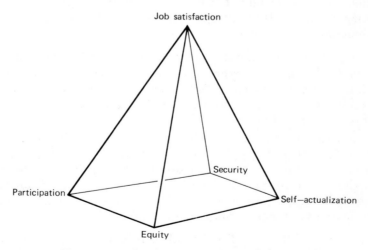

Figure 7.5 Maccoby's model of job satisfaction.

does is usually being done for the first time. People with uninformed opinions will resist or try to influence the process, since many people feel they are expert "amateur" psychologists. Argyris (1976) describe three of the problems faced by practitioners:

1. Experience with the practical situation is important; the practitioner must become completely familiar with the actual work environment, the workers, and every aspect of the work being done. Theoretical knowledge is no substitute for experience; no amount of education in school can make up for a lack of knowledge of the real life situation. Practical experience makes it possible to evaluate how the model must be adjusted to the particular type of work and what to expect from the suggested changes.

2. It is difficult to really "know" a theory completely. It is not sufficient to know the concepts well enough to talk about them; a deeper and more thorough understanding is required for applying models in practical ways. There is an analogy: if you show a sequence of snapshots of a person moving, then ask someone to repeat the series of movements in the pictures, it will probably prove impossible. If you show the same person a movie of the sequence of movements, better understanding will enable a more accurate repetition of the sequence. To apply a theory, according to the analogy, you must know all the factors involved (the snapshots) and the ways they are connected (seen in the movie).

3. The practitioner faces considerable insecurity. He or she must take the risks involved in doing something for the first time. In designing or redesigning work according to a particular theory, there is a chance for making highly visible mistakes. Only familiarity with the workers, tasks, and theory itself can instill confidence that, with or without mistakes, the new design will be a good one.

7.3.2   CHANGE STRATEGIES

Management can employ three different strategies in implementing changes in work. These are outlined below.

The Expert Strategy

Traditionally, when a problem exists in industry the management first consults with experts to determine the nature of the problem and how to solve it. Typically, the expert interviews many people, examines company records, and analyzes work methods. From this study a report is compiled, highlighting problems. The contents of this report are generally received by employees with considerable hostility and resentment. Defense mechanisms are aroused by the "invader," the expert brought in to investigate the workers and the work. Since the expert is chosen by and works closely with top levels of management, the conclusions of the report go straight to the top without consulting the people most directly concerned. Sometimes recommendations from such an expert can be turned into meaningful changes; more often, opposition from the employees makes this impossible. Often the report is paid for, filed, and forgotten.

There are some aspects of company operation that might benefit from study by an expert outsider. But analyzing employee attitudes and behavior is best done by someone close to the situation who can be effective without arousing resentment in the workers under study. Recommendations from such a study will probably be more informed and relevant as well.

The Education Strategy

To change behavior at work through education or training, the training should take place at work. In the workplace, employees get immediate feedback about how well they are doing. They also have opportunities to work with and learn from more experienced people. Training courses given outside of work have little value except to train people in basic task performance. For instance, factory courses on how to operate new equipment can be useful. Even this training, however, would probably be more effective done in the actual workplace where the equipment will be used.

The Participation Strategy

The best strategy for affecting attitudes, actions, job satisfaction, or productivity is to discuss issues openly at work. Since such discussions touch on the integrity and abilities of individuals, this strategy requires taking risks. Workers may feel threatened by open assessments of their efforts and capabilities. Therefore it is necessary that workers have confidence in the fairness and insight of the manager; without trust they will not be honest in discussions or committed to new goals. Managers must not be tempted to use this setting to manipulate people; such attempts are easy to see through and can cause irreparable damage. The discussions might start between managers and representatives of the workers, identifying problems and presenting coherent viewpoints on them. In some countries in Europe, this is even required by law. Most enlightened management people go well beyond the legal requirements in using this strategy because the results are so good.

7.4 GROUP WORK

So far we have discussed the design of individual tasks. We have also suggested that motivation and productivity might be increased if work could fulfill some of the workers' needs for social interactions, affiliation, and the like. Group work design may be the next step in increasing job satisfaction for many workers now doing individual jobs.

Often when people work together in groups, the individuals become more involved with the work and feel less alienated from the situation. In the construction industry, group work is the rule rather than the exception. So rather than argue the case for group work, we outline recent research findings on how such work should be organized. The studies were done in manufacturing industries; there have been no similar studies for construction work (see Chapter 8).

Gustavsen (1969) described some conditions for successful group work. He found that workers in a group were reasonably satisfied when the tasks they did were linked in some way; an especially appropriate way of linking people's jobs is job rotation. Good verbal communication between workers on the job was also found to be important for effective group work. Gustavsen found the highest level of job satisfaction among workers when all the group's tasks taken together constituted an identifiable process that clearly contributed to a meaningful product. He also pointed out that it is important for the group to have authority to make decisions concerning the work: Table 7.4 lists the most important areas of decision making. It seems that most of the conditions described by Gustavsen already exist in construction work.

TABLE 7.4 Work Group Decisions

1. When to work
2. How to work
3. What support activities (maintenance, administration, etc.) to perform and when to perform them
4. Work methods to use
5. How to divide work among members
6. Who is to be group leader, if necessary

7.4.1   THE ORGANIZATION OF GROUP WORK

A number of recommendations concerning how to organize group work emerged from studies at the Volvo factory in Gothenburg, Sweden (Almgren et al., 1979a,b). It is generally believed that designing work so that it can be done by relatively independent, autonomous groups increases workers' job satisfaction and productivity. It is very difficult, however, to reorganize work that is done as a series of individual tasks so that it becomes a group job. We do not describe the elaborate research done at the Volvo factory concerning methods of work design, but only give the conclusions regarding work in autonomous groups.

Attitude Change

Individual workers and management must realize that the work group is the planning and executive unit. This has important implications. The worker must change from doing as told to taking responsibility and initiative at work. Since it is demanding (as well as satisfying) to take this kind of responsibility, the company must buy responsible behavior from workers by paying with increased freedom at work. Unless the company takes this risk there will be little change in the amount of responsibility workers will accept. This whole process takes time; as much as 3 years may be required for behavior and attitudes to change in the desired way.

Recall another principle: the attitude change must start within the work group itself. And it must be supported by training in new ways to act on the job, new ways to interact with coworkers, and so on. A behavior change begins with the individual's willingness to learn and relearn as the work requirements change. Workers must be allowed to make a few mistakes without grave penalties in order to develop confidence in the new system and develop commitment to the new goals. Where a participation strategy can be used to implement the change, it is good to discuss with everyone involved the attitudes that will be the basis for future operation of the group. If opinion leaders can be identified, it is useful to get their cooperation from the start so they can lend support to the changes.

Management attitudes must change as well to allow the group the necessary freedom and responsibility to work effectively. Management can actively encourage the process by communications that support the changes and by positive feedback when the group is functioning well. Supervisors must be flexible enough to support the change in spite of what they may perceive as a lessening of their own authority.

Training

Two kinds of educational processes can support and facilitate the change from individual to group work. First, workers must learn the functions of roles of the entire operation with which the group will work. To become effective planners and decision makers, workers must know how all the jobs fit together to contribute to the end product. The second kind of education that is useful is some training in elementary group psychology. This will help group members understand their own reaction and defense mechanisms and to resolve interpersonal problems within the group. A little psychological training is also useful for workers involved in recruiting and training of new group members.

Management Approach

The role of management changes somewhat, from an emphasis on giving orders to an emphasis on providing the appropriate support for production. This includes timely delivery of materials to be used and provision of technical or economic expertise to help the group in solving problems.

Feedback

Evaluation of performance is essential for learning to take place quickly. Feedback must be given to the group soon after work is performed; evaluation of group performance given more than 8 hr later is not likely to have any effect. Once the group is competent to handle the work situation, it seems reasonable to give feedback on quality and quantity of production about once a week. It is essential to establish a realistic, agreed-upon production goal for the group. Evaluation should be given in terms of the established goal. Feedback should also demonstrate concern for workers and assure them that somebody "up there" is interested in them and paying attention.

Group Size

The optimum number of workers depends on the job and work conditions, of course. Where there is a choice, it has been found that groups of more than seven members tend to become fragmented and groups of less than four can become overly oriented to a single member.

Control Activities

It is a bad idea to have quality control or inspection work done by people outside the group. Work groups commonly protest such arrangements by lowering quality to keep the inspector busy. The group should have full responsibility for meeting quality goals.

7.4.2   RULES FOR GROUP WORK

To ensure a satisfactory level of productivity it is essential to establish agreed-upon rules for group work. Group members must understand quite clearly that to enjoy the freedom and variety of group work, each member must be willing to invest effort, loyalty, and time. The rules must be developed by consensus between group members and management, and it is vital that each participant understand the specific implications of each rule. The most important basic rules concern the following areas.

Production Objective

This is a clear statement of production goals agreed on by workers and management. There should also be a description of any working restrictions for attaining the goals and the level of product quality required.

Production Conditions

Rules should specify the physical prerequisites for production. It should be clear what equipment and materials are required for production; rules should also detail how problems such as material deficiency, cleaning, inventory, communication, and so on should be handled.

Work Performance and Worker Interactions

These rules should state work procedures and describe the individual's involvement and responsibility. An example is a rule concerning how to support co-workers in a way that furthers group goals (rather than ganging up on a slower worker or competing with co-workers). Job rotation is extremely valuable where practical; it forces each worker to learn all about the group's job and the situations specific to each of the tasks that make up the overall job.

Freedom

Rules are necessary to state the limits of freedom. There should be rules regarding permission to leave, overtime, flexibility of working hours, and rest periods. There should be a statement about what the group will do if the day's work is finished early.

Recruiting

Definite procedures concerning personnel turnover are needed to keep the group working as a closely integrated, cooperating unit. There should be clear statements concerning recruitment procedures, choice among applicants, training of new members, and working with replacement personnel.

Development of Competence

A newly created work group must understand that management expects increasing competence over time. A rule should set up a system of formal group discussions of performance and ways to work more effectively.

Structure

The group should not have a permanent foreman; this function should rotate among group members on, for instance, a monthly basis. The "foreman" is really a contact person who coordinates with other persons concerning purchasing, time reporting, materials deficiencies, orders for change, control, quality evaluation, budget, and so on.

7.4.3 SUMMARY

1. To be a practical tool, a theory must be relevant, general, and easy to communicate.
2. Using a theoretical model requires intellectual understanding, security and familiarity in the applied setting, flexibility of understanding, and practice.
3. Maccoby's motivation theory is an example of a practical one; Maccoby reduces job satisfaction to four essential components.

4. There are three strategies for implementing change in a work situation:

   a. The expert strategy is usually ineffective because it makes employees defensive and resistant to proposed changes.

   b. The training strategy has its value in introducing changes but usually does not lead to permanent change unless training is conducted at the work site.

   c. The participation strategy is generally the most powerful, since it involves the employees in the decision process. Workers more readily accept changes arrived at by consensus.

5. Group work may be more satisfying than individual work because it answers to more of the workers' needs.

6. Redesigning a work situation to go from individual to group work is difficult and requires time. Attitudes must change, training is required, and the new system must be tried and adapted over a long period of time.

7. Clear rules for group working help ensure satisfactory productivity. Rules must be arrived at by consensus in such areas as production goals and conditions, work performance and worker interactions, freedom, recruiting, competence, and structure.

REFERENCES

Almgren, B., Sjödin, B., and Söderberg, S. *Styrning och grupporganisation* (UARDA-rapport No. 4, 1979). Sweden: Division of Industrial Management, Chalmers University of Technology, 1979.

Almgren, B., Sjödin, B., and Söderberg, S. *Införande av Grupporganisation vid Volvo Lastvagnars Försöksanläggning i Arendal* (UARDA-rapport No. 6, 1979). Sweden: Division of Industrial Management, Chalmers University of Technology, 1979.

Argyris, C. Problems and New Directions for Industrial Psychology. In M.D. Dunnette (Ed.), *Handbook of Industrial and Organizational Psychology*. Chicago: Rand McNally College Publishing Office, 1976.

Friberg, M. "Är lönen det enda som sporrar oss att arbeta?" *Sociologisk Forskning*, **4,** 1975.

Gardell, B. *Produktionsteknik och arbetsglädje*. Stockholm: PA-rådet, 1971.

Gustavsen, B. *Bedriftsorganisasjon—alternative modeller*. Oslo: Tanum, 1969.

Herzberg, G., Mausner, B., and Snyderman, B. *The Motivation to Work*. New York: Wiley, 1959.

Klein, L. *New Forms of Work Organization*. London: Tavistock Institute of Human Relations, 1974.

Maccoby, M. Personal communication. 1978.

Maslow, A. *Motivation and Personality*. New York: Harper, 1954.

Rokeach, M. The Nature of Attitudes. In D. L. Sills (Ed.), *International Encyclopaedia of the Social Sciences*. New York: Macmillan, 1965.

Trist, E. and Bamforth, K. W. Some Social and Psychological Consequences of the Longvall Method of Coal-Getting. *Human Relations* 4(1), 3–38, 1951.

Vroom, V. H. *Work and Motivation*. New York: Wiley, 1964.

CHAPTER 8

THE ORGANIZATION OF WORK IN CONSTRUCTION*

RAYMOND E. LEVITT, Ph.D.

Department of Civil Engineering
Stanford University
Stanford, California

8.1 PURPOSE AND SCOPE

The intent of this chapter is to describe the organization of construction firms, projects, and crews and to highlight some of the ways in which their work is structured differently from work in other industries.

The theme of the chapter is to analyze how the organization of work in construction firms depends on characteristics of the product—constructed facilities—and on the characteristics of its resource markets. The close relationship between design practice, labor and materials markets, and work organization results in a considerable degree of inertia against change, and may help to explain why the industry has so often been accused of failing to innovate.

We begin the discussion by pointing out some unique aspects of the construction product, emphasizing the effect that these characteristics have on the overall organization of the industry's resources, participants, and markets. Subsequent sections look at the evolution of the structure of firms, projects, and work crews in response to the demands of construction products. We also devote considerable attention to the

* The author gratefully acknowledges the assistance of William H. Pearce in researching and compiling sections of this chapter.

structure and motivational aspects of work crews. The chapter con-
cludes with an analysis of performance evaluation in the industry.

8.2 THE NATURE OF THE CONSTRUCTION PRODUCT

The characteristics of constructed facilities have strong implications
for the organization and procedures of the entire construction industry.
Northrup and Foster (1975) observe the following:

> As in most industries, the salient characteristics of the construction in-
> dustry and its labor market are reflections of the nature of the construc-
> tion product. Although there is great variation in the kinds of facilities
> produced by the industry, all construction products exhibit a number of
> common properties from which devolve the structure and organization
> of the industry, and, derivatively, the problems associated with labor
> force adjustment.

Perhaps the most obvious characteristic of the construction product
is immobility. The product, such as a building, highway, dam, or factory
is normally a permanent structure that is produced and assembled on
location largely from raw materials (sand, gravel, cement) or primary
materials (steel bars, bricks). This on-site work in construction creates
certain considerations peculiar to the industry, which are in sharp
contrast with the usual conveyor-belt, mass production found in the
manufacturing industry. Each construction project presents unique
problems and requirements that must be dealt with individually. For
example, variations in climate or topography necessitate special solutions that cannot be formalized into standard, all-purpose designs,
even in product types with a common purpose. Therefore, the designer
as well as the contractors must take careful consideration of all the
specific requirements and limitations for each project.
   The industry must hence tailor its construction team for each individual building project. Just as the hospital now utilizes a different
team of medical specialists to perform each critical operation, so the
construction industry organizes a group of specialty designers and contractors to meet the needs of each construction project. This trade specialization provides the flexibility to expand or contract to the construction project requirements, and to alter the mix of craftsmen (e.g.,
carpenters, electricians, ironworkers) to the particular mix that is
needed to efficiently produce the desired product according to the
planned schedule.

This uniqueness of each construction project is frequently accompanied by a relatively high degree of technological sophistication. Today's refineries, power plants, and even buildings contain highly advanced components, such as power sources, process controls, climate controls, and emission controls. These components are difficult to assemble on site and must often be interconnected with other complex components. Logistic management (e.g., control of time and cost) and technical coordination are both critical problems which, combined with the ever-changing skill requirements at different stages in the construction process, require professional management personnel to ensure a fluid operation.

In the construction industry, the combination of raw materials and labor in the construction process frequently results in a costly product. Its unique and "contracted" nature discourages complete standardization. With the builder carrying the brunt of the financial burden and possessing only a relatively small amount of capital, a degree of risk is present that requires unusual financial procedures. Since the product is generally durable as well as costly, rework or replacement may sometimes be postponed during periods of unfavorable economic conditions. This introduces an enormous fluctuation in construction demand; this is perhaps the single most important determinant of the organization of the industry. Economic cycles that result in some fluctuation of demand for mass-produced goods such as automobiles create boom-or-bust conditions for the more volatile construction markets, especially home building, which fluctuates countercyclically with short-term interest rates. To cope with this demand fluctuation firm organizations, subcontracting procedures, and labor markets have had to evolve flexible structures.

Finally, the product is normally contracted by a sponsor of the project rather than initiated by the builder. From a marketing point of view, therefore, the construction industry is a type of service industry rather than a manufacturing one. There is no formal market for goods; rather the sponser approaches the industry to initiate the production of a desired product. This exacerbates the uncertainty of fluctuating demand facing firms in the construction industry and must be carefully considered in understanding the characteristics of the industry.

Thus it is evident that there are some characteristics that are peculiar only to the construction product, and that these present special requirements that must be reflected in the structure and organization of the construction industry. The unique, "custom-built," on-site nature of the product and its strong fluctuations in demand have implications for all facets of the industry. In the following section, we discuss these industry characteristics.

8.3   THE NATURE OF THE INDUSTRY

In the preceding section, we investigated the nature of the product that is produced in the construction industry. In this section we continue this investigation by examining the influence of the product on the nature of the industry. After a general overview of construction, we discuss the resource markets (material, equipment, finance, and labor) that exist within the industry and the firms that transform the raw materials into the desired product.

8.3.1.   COMPOSITION OF THE INDUSTRY BY PRODUCT

The construction industry may be conveniently divided into two sectors, public and private. Approximately 25% by value of construction is considered public, with 20% initiated by state and local government (with federal aid) and 5% fully funded by the federal government. As seen in Table 8.1, the private sector may be further subdivided into residential and nonresidential construction.

Residential construction is sometimes referred to as the home-building industry. It includes the construction of multifamily as well as single-family dwellings. The nonresidential sector includes commerical construction (e.g., office buildings and stores), industrial work (factories), and utilities (power plants). Highway construction is exclusively restricted to the public sector. Heavy construction (dams, airports, subways) is found in both the private and public subdivisions. These divisions in the construction industry will again be of interest when we examine the labor market.

8.3.2.   PARTICIPANTS IN THE INDUSTRY

The construction industry consists of the owner/investors, the architects/engineers, contractors (both general and specialized), material and equipment suppliers, financiers, and labor force.

Architects/Engineers

The architects and engineers in design and consulting firms are contracted by the owner to assist him in the initiation and supervision of

TABLE 8.1 Nature of Construction Activities, 1972[a]

Private construction 75%		Public construction 25%	
Residential 42.0%	Nonresidential 33.0%	State/Local 20.0%	Federal 5.0%
New dwellings 35.0%	Commercial	Buildings 9.0%	Buildings 1.0%
	Office buildings 11.0%	Highways/streets 9.0%	Highways/streets 1.0%
Additions/alterations 5.0%	Stores/shopping centers 4.5%	Conservation and other 2.0%	Conservation 2.0%
	Garages/warehouses 4.5%		
	2.0%		
Non-housekeeping 2.0%	Industrial 5.0%		Military facilities 1.0%
	Utilities 10.0%		
	Other 7.0%		

[a] U.S. Department of Commerce (1972).

221

the project. Their firms consist of employees who are assigned to specific projects. Their product is a set of drawings that are developed so that they may be disaggregated down to the level of each specialty subcontractor for his independent use. Thus the designer recognizes the inherent complexity of the construction product and the requirement for individual experts to perform specialized tasks. This type of service, of course, is physically more mobile than the constructed facilities that are its end product. As such, they can easily aggregate regional fluctuations in demand by serving national markets. Consequently, their work force does not experience the large fluctuation that is evident in the blue-collar labor force.

Material and Equipment Suppliers

The material and equipment suppliers provide the raw materials and machinery that the contractor requires in his construction process. Although normally considered outside the construction industry, their importance in the construction process is obvious. The prices for their materials and equipment are subject to the normal variations present in the entire U.S. economy. The continual increase in the cost of construction materials, such as lumber and steel, has had clear implications for the cost of construction, and with these increases comes a renewed interest in cutting costs through better management or utilization of alternative materials.

Financiers

The financial procedures that exist in construction are unique to that industry. The contractor generally has only a small amount of capital, and hence the funds for costs incurred during the construction project must be obtained elsewhere.

Construction financing consists of short-term and long-term borrowing. Short-term financing is provided in the form of construction loans. These loans are provided by conventional financial institutions as listed in Table 8.2.

As shown, the majority of short-term loans are provided by commerical banks and savings and loan associations (51.8 and 31.6%, respectively). The construction loan provides the required financing during the construction process. Other forms of short-term borrowing, such as gap-financing and front money, are used when required. Gap financing is a loan based on the difference between the economic worth and the replacement cost of the structure. A lender will provide a cer-

TABLE 8.2   Construction Loans by Major
Construction Lenders in 1968[a]

Lending institution	% of total dollars in construction loans attributed to each lender
Commercial banks	51.8
Savings and loan associations	31.6
Mortgage bankers	8.8
Mutual savings banks	4.7
Life insurance companies	2.1
Real estate investment trusts	1.0
Total	100.0

[a] After Schulkin (1970).

tain percentage of this difference after a construction loan is obtained. Front money is a loan that is acquired to pay for costs incurred prior to receiving a construction loan.

For long-term or permanent financing a mortgage loan is required. Generally a commitment for mortgage must be acquired before receiving a short-term loan. The mortgage loan is generally provided by savings and loan associations and life insurance companies; see Table 8.3.

When money becomes tight and short-term interest rates rise above mortgage rates, the supply of money to savings and loan associations and other long-term lenders dries up very quickly. As available capital decreases, construction declines and many firms are forced to close down. This is particularly true for builders of residential properties, whose major financiers are savings and loan institutions. Recent measures have been taken by the federal government to encourage lenders to provide funds to construction buyers during periods of tight money conditions. But residential construction volume is still very closely related (inversely) to short-term interest rates, which depend on the discount rates charged by the Federal Reserve Bank.

Since industrial and commercial construction are funded by private investors as well as financial institutions, their dependence on interest rates is not as critical. Heavy and highway construction, on the other hand, are funded almost exclusively by government agencies and are often used to provide countercyclical fiscal stimulus to a country's economy.

Thus, because of the industry's heavy dependence on outside fi-

TABLE 8.3 Mortgage Loans Outstanding by Type of Lender and Type of Property In Percent, Year-end 1969[a]

Lending institution	Farm Properties	Commercial Properties	Multifamily	One- to Four-Family	Lending institution's % of the grand total
Savings and loan associations	—	2.5	2.7	27.8	33.0
Mutual savings banks	—	1.7	2.9	8.5	13.2
Commercial banks	1.0	5.2	0.8	9.8	16.7
Life insurance companies	1.4	6.0	3.0	6.6	17.0
All others[b]	4.6	2.6	2.8	10.0	20.1
Property's % of the grand total	7.0	18.0	12.2	62.8	100

[a] SOURCE: U.S. Savings and Loan League (1969).
[b] This source defines all others to include federal agencies (FNMA, FHA, VA, — 6.4% of the grand total), trust departments of commercial banks, pension funds, nonprofit institutions, credit unions, real estate companies, and individuals.

nancing, all its sectors are influenced—in different ways—by trends in interest rates, money supply, and overall economic strength of the larger economy of its country.

Labor

The "contracted" nature of construction work results in a large degree of market uncertainty within the industry. This uncertainty also makes the character of the labor market unique in several ways. Unlike most other industries, there is no strong employer-employee relationship for most workers. Rather, the employer generally hires his workers only for the duration of the project. Thus a floating labor force exists which moves from demand to demand (i.e., project to project). The contractor, on the other hand, is freed from maintaining his required working capital, and increases his flexibility to respond to demand fluctuations.

Obviously the worker faces a large degree of job insecurity. This often results in a lack of loyalty between employer and employee. However, the unions that exist in the U.S. industry attempt to provide job security. There are 15 building trade unions involved in construction which cover about 30 distinct occupational groups. The 15 AFL-CIO unions are listed in Table 8.4. Many of these unions are not exclusively involved in construction (e.g., only 19% of electrical workers are in construction).

Early in their development, construction unions established exclusive jurisdiction for their special trade. This means that any work that requires their special skill can be performed only by members of their union—for example, only electrical workers can perform electrical work. These jurisdictional boundaries are constantly being negotiated between the various building trade unions. Special industry panels have long been used to mediate jurisdictional disputes between two or more trades. With technological changes, task distinctions that were previously established become unclear. For example, when metal studs replaced 2 × 4 in. wooden studs for framing in construction, these could have been claimed by the carpenters, the sheet-metal workers, and the ironworkers, based on the way in which their respective jurisdictions were defined. Since jurisdictional disputes may result in strikes, and thus work stoppages, it is important to arbitrate and settle any such differences as fast as possible. Owing to the existence of industry panels that publish their findings in the so-called "Green Book," (Building & Construction Trades Dept., AFL-CIO, 1973) and to con-

TABLE 8.4   AFL-CIO Building Trades Unions, 1971[a]

Union	Total membership[b]	Percent in construction	Number of locals
Asbestos workers	18	99	121
Boilermakers	138	65[c]	425
Bricklayers	143	100	862
Carpenters	820	80[c]	2435
Electrical workers (IBEW)	922	19	1677
Elevator contractors	17	100	109
Granite cutters[e]	3	80	23
Ironworkers	178	61	320
Laborers	580	79	900
Lathers	15	100[c]	289
Marble polishers	8	80[c]	123
Operating engineers	393	65[c]	279
Painters	210	77	1000
Plasterers[d]	68	99	500
Plumbers	312	80[c]	680
Roofers	24	100	209
Sheet-metal workers	120	60	n.a.

[a] U.S. Bureau of Labor Statistics, *Bulletin 1750*, 1972.
[b] Numbers in thousands.
[c] Estimated
[d] Merged into the Carpenters, 1979.
[e] Merged into the Marble Polishers, 1980.
n.a.: not available.

siderable nonunion competition in many parts of the industry, juris-dictional work stoppages are now infrequent.

An enormous number of local agreements are entered into each year between local building trade unions and contractors or local contractor associations, in which union and management representatives engage in collective bargaining negotiations. Union strategies include having contract expiration occur during the busy summer work season, lob-bying for unemployment benefits to striking workers in state and fed-eral governments (to facilitate strikes), and utilizing strikes or pick-eting when necessary. Management, on the other hand, attempts to organize associations for bargaining or strikes and to reduce labor with automation or more powerful machinery. In the recent past, construc-tion unions have negotiated with considerable success. Union wages, including benefits, are relatively high compared to manufacturing wages. Part of this differential can be construed as compensation to construction workers for their lack of job security and high levels of unemployment. As a result, however, contractors have looked toward increased utilization of nonunion labor, as well as better management techniques, to reduce the already high construction costs.

Open-shop, or nonunion, construction has increased as a result of these attempts to reduce costs. A recent survey (Bourdon and Levitt, 1980) revealed that nonunion labor is moving into previously union strong holds. Nonunion construction is no longer restricted to its traditional sector of small residential construction. It has expanded to compete with union labor in such fields as small heavy construction and large manufacturing construction. At the same time, nonunion strength is still present in rural areas and in the southern states.

Nonunion wages are normally below local union wages. Yet this savings is somewhat reduced by the fact that the more unskilled open-shop laborer requires more supervision. This was emphasized in the study by Bourdon and Levitt (1980), which showed that per 10 nonunion journeymen there are five leadmen/foremen and nine helpers. For 10 union journeymen, however, there are only one foreman and two apprentices. Thus, although nonunion construction replaces some journeymen with lower-paid helpers, more higher-paid supervisors are also required.

A brief mention of the Davis–Bacon Act is in order. This federal regulation requires that workers on federally funded projects be paid the local "prevailing" wage as determined by the Department of Labor. This stipulation often results in a wage close to the union rate. This act has its greatest effect on highway and housing construction.

The relative efficiencies of union versus nonunion labor were found by Bourdon and Levitt (1980) to be most strongly related to project scale. Small projects require broadly trained, flexible workers to carry out a variety of tasks. The union jurisdictional boundaries are too narrow here; if enforced, they result in standby labor. Very large projects, on the other hand, permit specialization to be efficiently extended far beyond the 15 trades mentioned in Table 8.4. For example, on a major refinery project it is quite feasible to keep a worker busy full time merely cutting and threading pipe. Clearly, from the management point of view, a 4-year apprenticed union plumber is overtrained—and overpaid—to do this work. Consequently, nonunion contractors are moving aggressively into the superprojects with modular training packages to teach previously unskilled workers narrow skills.

In the medium-size range, however—projects employing 50 to 500 on-site workers—the union occupational breadth appears to be about right. On these projects (high-rise buildings, heavy construction projects) union firms are found to dominate. Their local pools of skilled labor, the hiring halls, provide union contractors in this size range with the additional advantage of being able to easily assemble and then lay off a large, trained work-force. The capacity of the local hiring halls is, however, insufficient to man superprojects; hence this benefit is not really shared by union contractors on major power-plants or chemical process plants.

Although there have been problems in the past with corrupt hiring hall procedures, and with discrimination against minorities (See Chapter 10), the hiring hall does provide union contractors with an easy means for labor recruitment. Agreements between a contractor and the local-union often include the stipulation that workers can be rejected by the contractor. In nonunion construction, there has been an increase in the use of contractor-organized employment agencies, but the majority of workers are still hired through employee referrals, continual informal contact, or newspaper advertisements. State employment services and vocational schools are additional sources of manpower.

The union participates in training for their skilled labor through formal apprenticeship training programs. These ensure their level of technical competence but limit the number of skilled craftsmen. This skill level is especially critical in the more technical trades, such as electricians and plumbers (see Table 8.5). Nonunion constructors are attempting to provide formal training through their own training programs, but government regulations that establish apprenticeship requirements have hampered their participation in formal apprentice training.

Management in the construction industry is organized into contractor associations. These organizations attempt to assist contractors in their negotiations with labor organizations, and provide lobbying at the state and national levels. The Association of General Contractors (AGC) includes about 10,000 general contractors with approximately 120 chapters. About 40% of its members are open-shop contractors. The National Constructors Association (NCA) is concerned with unionized industrial and heavy construction. The Associated Builders and Contractors (ABC) is the fastest growing organization, and includes some 10,000 members and 50 chapters. It consists of mostly open-shop, small, general, or specialty firms. Other associations are organized on the basis of branch of construction, such as the National Association of Homebuilders (NAHB).

TABLE 8.5   The Percentage of Each Craft that
Learned the Trade through Apprenticeship,
Technical School Training, or Training in the
Armed Services[a]

Construction craftsman	% that learned trade through formal training
Bricklayers, stonemasons, and tilers	44.7
Carpenters	31.1
Electricians	72.9
Excavating, grading, and road-building machine operators	11.2
Painters	27.8
Plumbers and pipe fitters	55.0
Tinsmiths, coppersmiths, and sheet-metal workers	70.9
Cranemen, derrickmen, and hoistmen	17.5
All construction craftsmen	39.4

[a] SOURCE: Mills (1972).

Contractor Firms

The contractor firms in the construction industry vary in both size and
purpose. The general contractor firm may include basic trades, such
as carpenters, ironworkers, cement masons, and laborers, or may provide only the management and supervision, subcontracting all labor
to other contractors. Most firms are relatively small and require little

TABLE 8.6   Distribution of Firms and Employment
by Company Size[a]

Construction company size (No. of employees)	Percent of companies	Percent of labor force employed (approximate)
≤4	82	25
5–9	8.6	10
10–49	8.0	30
≤50	1.4	35

[a] SOURCE: U.S. Dept. of Commerce, Bureau of Census, 1972
(1977 Census Data unavailable as of June, 1979).

capital for their employee payroll. Approximately 80% of all construction firms are sole proprietorships, 15% are corporations, and 5% are partnerships (U.S. Bureau of the Census, 1972) See Table 8.6.

Specialty firms that perform a specific trade, such as electrical work, provide a solution to the problem of maintaining flexibility in an uncertain market that produces unique products. They provide the general contractor with the expertise to perform complex or technical tasks, without the burden of being on the contractor's payroll when their work is not required. A contractor can thus more easily adjust to changes in demand that cannot clearly be foreseen.

8.3.3 RECAPITULATION

In this section, we have provided an overview of the manner in which the industry has evolved to properly adjust to the demands made by the construction product. Labor has organized into trade unions which establish exclusive jurisdiction over their trades, preserving their workers' jobs and providing a pool of skilled craftsmen to specialized firms, to perform their part in the construction of complex products. The general contractor subcontracts his work to these firms and provides the management, and sometimes the basic skills, for creating the construction project. Materials and machinery are supplied to the contractors in standard measurements as determined by the industry and government-regulated building codes and specifications. The designer produces drawings and specifications set up in a way that permits separate contracts to be let to a large number of specialized construction firms. Thus the industry has adapted to the requirements dictated by the nature of its product.

8.4 ORGANIZATION OF FIRMS IN THE INDUSTRY

The previous section provided an overview of the organization of the U.S. construction industry. We now focus our attention on the subunits within the industry by investigating the internal organization of firms. A framework of organization theory and a discussion of alternate organizational forms are provided first. Firms in the various sectors of the industry are then analyzed using this framework, and their most common organizational forms are described.

8.4.1 ORGANIZATION THEORY FRAMEWORK

The organization design problem has been succinctly summarized by Galbraith (1973) as a selection of the proper balance between the benefits of specialization and the costs of coordination:

> That is, the greater the degree of specialization, the greater is the effectiveness of subtask performance. But the greater the degree of specialization, the greater is the difficulty of integrating the subtasks into successful performance of the entire task.

Clearly the construction industry requires a coordinated effort of specialized skills, although the number of diverse tasks varies from project to project within a sector of the industry, as well as varying between these sectors (e.g., power plant construction involves greater complexity and a larger collection of skills than the relatively simpler process of residential construction). This variation poses different requirements of specialization and coordination, and firms adapt to these differing requirements by selecting different organizational structures.

Thompson (1972) depicts an organization as a production ("technical") core, surrounded by "boundary-spanning units" which buffer the core from the effects of uncertainties in the organization's environment. Where the degree of environmental uncertainty is high—for example, in construction—these boundary-spanning units play a very significant role in the functioning, and survival of the organization. Furthermore, Galbraith (1973) suggests that as uncertainty increases, the organization's communication channels may become overloaded, leading to poor performance.

Thus construction firms must develop structures to cope with the strong need for specialization and the high levels of environmental uncertainty that characterize its production process.

8.4.2 ALTERNATIVE ORGANIZATIONAL STRUCTURES

There are basically three distinct types of organizational structures that may be employed by firms in the construction industry. Each structure possesses a different mode of coordination of participants within the firm, as well as a different allocation of authority and responsibility between product or project managers and functional division heads.

Classical Structure

The traditional approach to the design of an organization is the basic hierarchical structure depicted in Figure 8.1. It consists of a functional organization arranged in the standard pyramid configuration, and employs the principles of classical management theory. Subunits are grouped together by common tasks to achieve specialization efficiencies. Yet they can share their resources and provide mutual support. A central administration provides common services for the divisions. The functional specialization and the sharing of resources provides an efficient structure and results in a low amount of duplicated effort. However, it deemphasizes the importance of cooperation between specialist workers. In intensive technologies such as construction there are strong interdependencies between functional groups. These are not easily coordinated in a functional structure. The technical performance of each functional division becomes more important and immediate to the employee than the success of the overall project.

Project Organization

The other extreme in organizational approach is the project organization. Here employee loyalty shifts from the functional division, which

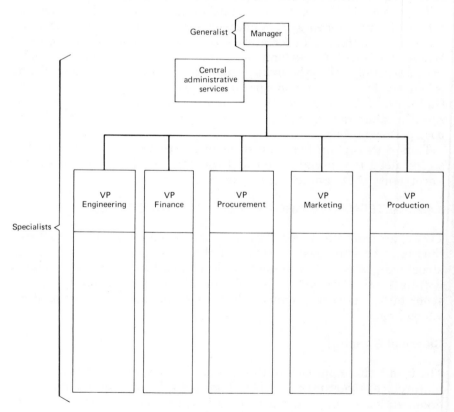

Figure 8.1 Functional organization.

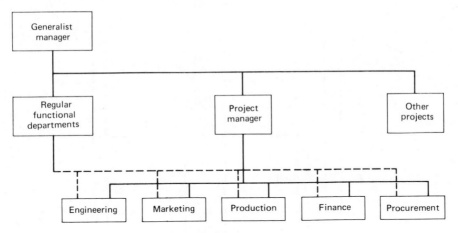

Figure 8.2   Project organization.

was paramount in the functional organization, to the project. A temporary organization is created which concentrates all its resources toward the project objective (Figure 8.2). The project becomes a self-contained unit with a project manager who possesses full authority over all his personnel. The employees are physically located with the project organization rather than with their functional departments. Individuals are usually delegated a certain amount of decision-making authority in committing the resources of the company. For example, a project manager may be authorized to purchase equipment up to $10,000, or to approve change orders or negotiations within specified limits. Compared to the functional organization, the project organization provides a more concentrated and unified effort under a single authoritative person. The disadvantages of the project organization are that it may upset the regular organization of the firm, and may duplicate subunits when several projects are in progress simultaneously. In addition, the project personnel may fear being laid off when the project is completed. For this type of organization to be efficient, the cost of forming a separate unit of project personnel must be less than the cost of maintaining a coordinated effort between the functional departments.

Matrix Organization

The matrix organization combines the approach of both the traditional structure and the project organization; see Figure 8.3.

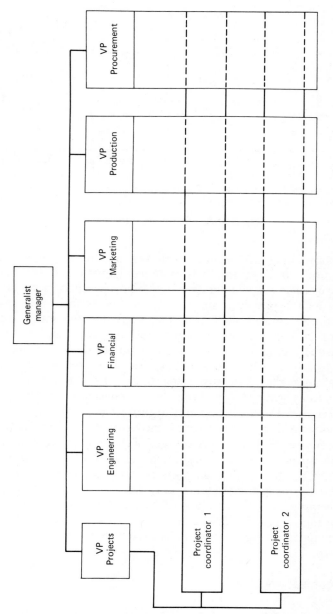

Figure 8.3 Matrix organization.

It may vary the emphasis in this combination to provide a structure which best fits the needs of the firm. Youker (1976) depicts the matrix as a continuum, centered between the functional or traditional approach and the project organization (see Figure 8.4). He further states:

> The dividing line between functional and matrix is when some individual is appointed with part-time responsibility for coordination across functional department lines. . . . Ordinarily, there is a clear distinction between a strong matrix in which most of the work is still being performed in the functional departments contrasted with a project organization where the majority of the personnel are on the project.

The various degrees of matrix organization are determined by the amount of coordination that is present. The weak matrix, closely resembling the functional structure, has only a part-time coordinator, whereas the strong matrix organization has its own project office.

The matrix structure thus requires a combined effort by both functional divisions and project heads. The employees work for both managers, and the firm must establish the amount of authority each manager possesses to delegate distinct responsibilities. The project manager usually determines *what* is to be done *when*, whereas the department

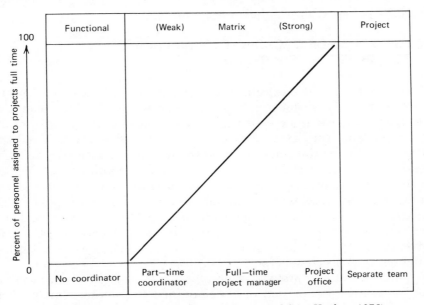

Figure 8.4 Organizational continuum (adapted from Youker, 1976).

chief is concerned with *how* it is to be done. In fact these distinctions are not always clear, so that close cooperation between these two managers is essential to preclude any conflicts. Such conflicts could seriously jeopardize the success of the project, with the individual worker torn between the authority of his department head and the authority of his project manager.

The various positions available in the matrix organization continuum provide a wide degree of selection and flexibility for the firm. The firm may utilize any combination of structures within its organization, or even throughout the various levels of the project. The matrix organization is especially useful when a number of projects are conducted at the same time.

The selection process must also consider the skills of both the project managers and the department chiefs. The project organization requires a knowledgeable leader with sufficient managerial and technical skills to successfully operate independently from the firm. The functional organization, on the other hand, places more responsibility on department chiefs and sophisticated planning and reporting procedures to ensure smooth operation of company projects. The matrix organization requires a diplomatic project manager who can effectively communicate with his peers, instill team loyalty in his employees, and anticipate conflicts that may develop. His authority is largely informal, and the potential for conflict is therefore high.

8.4.3 MODES OF COORDINATION FOR HIGHLY INTERDEPENDENT FUNCTIONS

Coordination is obviously important to ensure the successful, smooth operation of the various function in an organization. The amount of coordination that is necessary largely depends on the amount of interdependence that exists between interacting organizations and the degree of uncertainty associated with the performance of individual tasks. In a stable environment with few uncertainties, established rules and regulations coupled with hierarchical planning and scheduling may be sufficient to coordinate activities. But the unpredictable nature of the construction industry and the strong interdependence between specialty groups usually require more elaborate methods to ensure effective coordination both within functional groups (vertically) and between specialists (horizontally).

Galbraith (1975) provides a progression of more involved methods of horizontal coordination in his discussion of program management (see Figure 8.5). He begins with two processes that utilize the func-

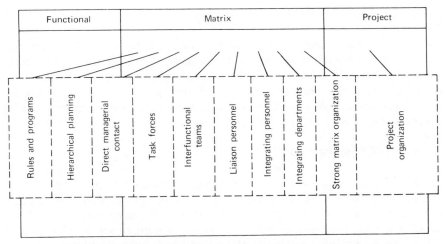

Figure 8.5 Coordination mechanisms in different organization types.

tional hierarchy and then proceeds to those that increase horizontal contact within the organization. Rules and programs specify interdependent behavior in simple situations without uncertainty. Hierarchical planning dictates the specification of outputs and goals rather than specific behavior. Exceptions are handled by the upper levels of the hierarchy. When the number of exceptions reaches a level where the flow begins to overload the organization, more advanced methods must be used. Direct managerial contact at lower levels encourages a more coordinated effort. Task forces may be formed as a temporary organization to solve coordination problems. When the problems are recurrent, more permanent interfunctional teams are organized. Coordination may increase to a level where the use of liaison personnel or integrating personnel or departments (with increased influence) becomes desirable. Finally, the organization may be structured in a matrix configuration or in a full project structure when the level of coordination is especially demanding.

The project organizations that exist within the industry include such a diverse collection of tasks and specialists that they may require several of these coordination mechanisms simultaneously. Periodic coordination may be facilitated by scheduling weekly job meetings. Informal, lateral communication, similar to the authorization of direct managerial contact mentioned above, is frequently encouraged within the organization to assist information flow and preclude potential conflicts.

8.4.4 TYPES OF FIRMS

As mentioned previously, the construction industry consists of a wide variety of firms with diverse objectives. These firms differ in both size and function. They include architectural design firms, construction firms (both general and specialized), consulting firms, and owner-agencies. They employ different organizational structures to achieve maximum benefits at a minimum cost.

Architect–Engineer Firms

Most architect–engineer firms start out as small partnerships of designers, who operate as generalists, with assistance from a pool of junior architects/engineers and draftsmen. Recently, however, several factors have produced economies of scale for design firms, so that some firms are becoming larger. At the same time increased pressures to limit liability have induced them to adopt corporate legal forms.

Figure 8.6 depicts a typical hierarchical organization for a large architect–engineer (A–E) firm. These firms have evolved as a combination of pure hierarchical and matrix structures (Logcher and Levitt, 1978).

The top line consists of a corporate headquarters and a staff that

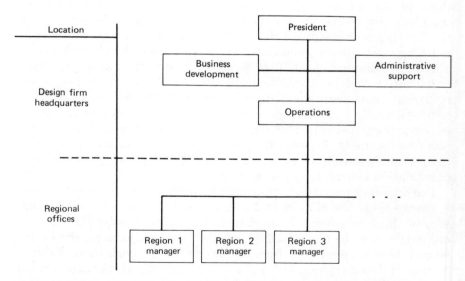

Figure 8.6 Hierarchical structure of a large architect–engineer firm (SOURCE: Logcher and Levitt, 1978).

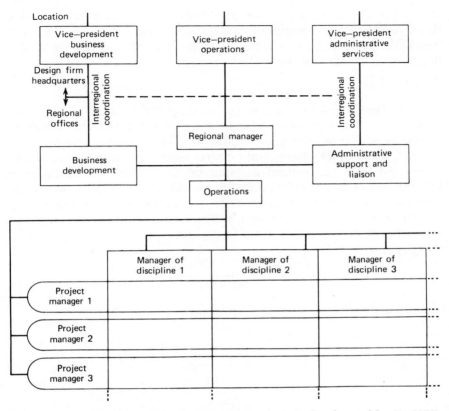

Figure 8.7 Matrix structure for project execution (SOURCE: Logcher and Levitt, 1978).

performs company administration for the entire firm (e.g., personnel actions, accounting, data processing, marketing). The next level includes the regional offices, which are generally dispersed from the corporate center. Their remote location and varying project conditions require a large degree of independent action. Thus the regional office acts as a separate profit center. controlling its own costs and revenues. It is organized as a second-level matrix structure, with specialists grouped by discipline while assigned to one or more current projects (see Figure 8.7). The department chiefs who head each discipline control the production costs and manpower utilization within their departments, and are evaluated in terms of budget constraints, manpower utilization, and technical quality. The project manager allocates the total direct cost of the project, plans and schedules the work flow, and negotiates additional costs. His or her performance is measured in terms or on-time completion and overall design efficiency. The success of the project clearly depends on the smooth cooperation of the department heads and the project manager.

Construction Firms

Construction firms can be general or specialized, large or small organizations. The structure of the firm is influenced by factors such as the tasks performed, the uncertainties that exist in its operations, and market, legal, and institutional requirements. In order to avoid delays associated with centralized decision making, much of the decision making might be decentralized to the building site. This is especially true for projects of high uncertainty, for example, underground construction or marine construction, where virtually all decision-making authority must be decentralized to the site.

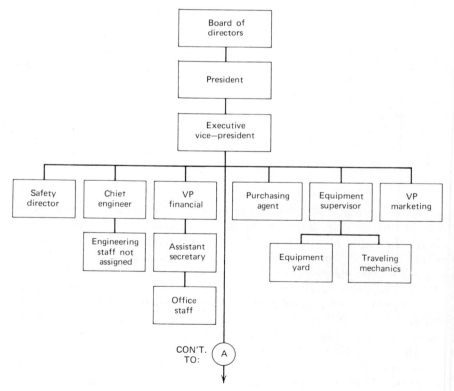

Figure 8.8 Medium-size heavy construction company.

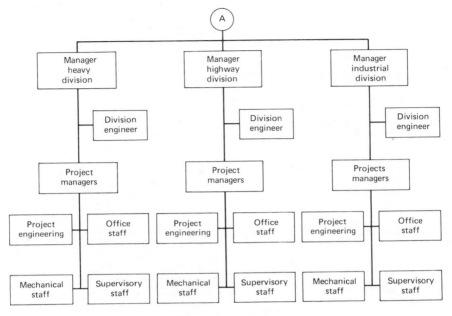

Figure 8.8 (continued)

Figure 8.8 provides an example of a medium-size heavy construction company where most of the decision making is decentralized. The company specializes in projects such as tunnels, locks, and dams that involve large technical and financial risks. The home office staff provides centralized services for the entire company, such as marketing and accounting. In addition, specialized and costly equipment such as cranes can be allocated to projects from the company pool of resources. The centralization of these functions reduces the need for resources and employees, which in turn decreases overhead costs.

The second level is subdivided into three functional divisions, with four functions included in subunits under the direction of the project manager. The divisions are separate profit centers that function independently. The benefits of this decentralization—increased decision-making ability at the lower levels—balance the additional costs incurred with the duplication of some services.

Owner-Agency

The owner-agency construction department or division provides construction services for the parent organization. Figure 8.9 provides an

242

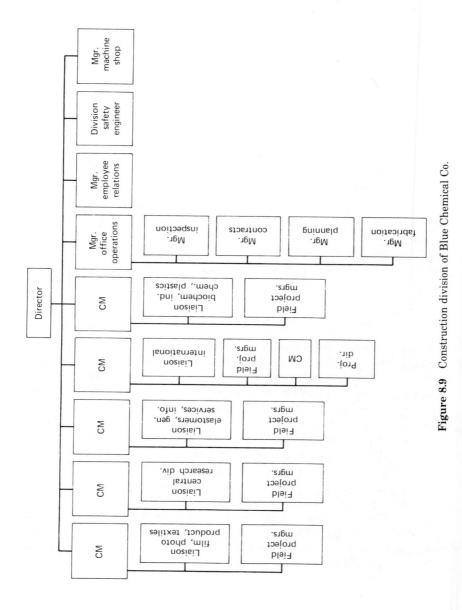

Figure 8.9 Construction division of Blue Chemical Co.

example of the organization of the construction division of the Blue firm, an international chemical corporation.

This type of construction division offers several types of advantages. As a part of the larger organization, coordination requirements and uncertainties that are present when dealing with outside organizations are virtually eliminated. As an in-house operation, it permits greater control of the project by the owner. The degree of control and participation that is desired by the owner depends on the nature of the project as well as the capabilities of the organization. In addition to providing ease of owner input regarding operating criteria and cost and schedule constraints, the construction division can also offer an early opportunity for company personnel to acquaint themself with the new facilities. This makes the future operation and maintenance of the facilities easier. The utilization of an owner-agency approach requires a full-time construction team which produces some overhead costs for the company. Thus it is important to provide a sufficient number of projects to keep the construction unit gainfully employed so that additional expenses are justified. However, to gain the full advantage of the owner-agency approach the organization must possess the in-house expertise to provide a meaningful contribution to the construction process. An alternative to the owner-agency approach is the use of consultants to aid the owner in achieving the desired project, when the owner lacks in-house capabilities to perform the construction or desires to limit his participation in the project.

Examples of these two different approaches are provided in two recent subway construction projects in the United States. The Metro Authority in Washington, D.C. (WMATA) used a number of in-house coordinators to manage technical and logistic aspects of design and construction; the Bay Area Rapid Transit (BART) in San Francisco delegated much of this authority to its general consultants. The results support the idea that a trade-off must be made between speed of decision making and dgree of owner control over project scope, cost, and schedule. The WMATA organization was slow to make decisions but retained tight control over decision making. This resulted in some delay costs. The BART organization, on the other hand, relied heavily on their consultants to make most decisions. The decisions were thus made quickly and centrally, but the owner perceived a considerable loss of control over decision making. The technical performance problems that have resulted from the very innovative design of BART may be partly attributable to this reduced owner involvement in decision making.

8.4.5   Decision Making at the Project Level

The more geographically dispersed a firm, the higher the need for independent project operations. Thus the project manager must be able to make proper decisions that are required during the course of the project when referral to higher headquarters would create costly delays. Those firms that require dispersed work forces, such as highway construction, utilize lower-level decision making to avoid lengthy communication networks and long delay times. The project manager is usually provided with the authority to make purchases and change orders or negotiations within the limits established by the company. The assignment of responsibility to the department heads and the project manager by top level management should be clear and definite. Depending on the type of organization (functional, matrix, or project) different delegations of responsibilities and authority for decision making are necessary. The extent of uncertainty concerning the project and the environment (such as weather and soil conditions) can also influence the delegation of decision-making authority within the organization. The decision that is made at the project level should be as good as the decision that would have been made at the home office if more time were available. Obviously the project manager sometimes has only partial information available. It is therefore important to communicate regularly with im so that the individual possesses the desired knowledge when the situation demands independent action.

Decision making for support functions that require only periodic information flow, such as marketing and accounting, generally can be located at the home office. Also, sharing of company equipment resources might involve decisions that must be made at a central level.

We have examined the three major forms of organization and the options that are employed by firms in the various sectors of the construction industry. We now turn to a discussion of the roles of members of these firms on projects and the structure of project organizations.

8.5   INTERACTION OF FIRMS ON PROJECTS: ALTERNATIVE PROJECT DELIVERY PLANS

A construction project requires the interaction of a large number of individuals from several distinct firms to construct a unique facility in a limited time. This requires a great deal of coordination and communication. The project organization must be selected after a careful evaluation of the needs and desires of the owner. Furthermore, clear distinctions must be made between the individual roles of the participants in the construction process to ensure that role conflicts are avoided. In Section 8.3 we briefly outlined the participants in the construction process. Here we perform a more detailed investigation of each individual's role through presentation of the alternative sets of contractual relationships (project delivery plans) that are commonly used in the industry.

8.5.1   THE PARTICIPANTS

The owner initiates the construction project. He may be an individual, an association, an agent of the government, or any other individual or group desiring the construction of a project. He may or may not be the eventual user or operator of the resultant product. He usually hires a consultant to provide professional advice on the design and other construction requirements. The degree of owner participation in the project after the initial design process is largely determined by the project delivery system he has selected.

The architect provides the design which satisfies the desires of the owner and supervises its construction to ensure that it is built according to his design. In addition, he may perform other functions, such as aiding the owner in the selection of a contractor. Recently, however, the role of the architect has become more limited to strictly the design responsibility.

The engineer consultant is either hired by the architect or hired directly by the owner. His role is especially important in industrial, heavy, and highway construction. He provides professional advice concerning the engineering aspects of the project, including foundation, structural, and mechanical considerations. Frequently he is permanently associated with the architect in a combined architectural–engineering firm.

The general contractor acts as the resource manager during construction. He may or may not perform some of the construction work himself. He subcontracts specialized tasks to other firms and supervises and coordinates their work. He is sometimes not included in the prior planning performed by the owner, designer, and engineer consultant. As a result design-construction problems often occur, resulting in confrontations between the architect and the contractor. Some recently introduced project delivery systems introduce the contractor into the construction process earlier to avoid this antagonism.

The specialty trade contractor provides the specialized workers, equipment, and know-how required to perform the many facets of the construction project. As discussed previously, this provides a means to maintain flexibility in an industry that has both a fluctuating demand and a unique product.

This is true because the demand for electrical specialty work in a region is an aggregation of the demand for electrical work on all projects in that region. As a result, the electrical (and other) specialty firms are able to keep their labor and equipment more evenly utilized than a general contractor could, by moving them from project to project, and by doing maintenance and repair work in the winter. Specialty trade firms are generally subcontracted by the general contractor, but in some project delivery systems may be hired directly by the owner.

Construction managers (CM) in the construction process have become more widely used in recent years. Many reasons have been suggested for this increase: the need to decrease project durations by overlapping design with construction in order to combat rising inflation; the greater coordination demands present in large projects; and the void left by the architect's departure from the detailed supervision and inspection phases of the construction process.

The duties of a CM are not yet adequately defined. The American Society of Civil Engineers (ASCE) defines a professional construction manager simply as one who specializes in construction management. This construction management is involved in the project planning, design, and construction phases. A general contractor may be considered a CM when he performs the management function in this three-man team, even though he may still perform some construction with his own labor force (Barrie and Paulson, 1976).

The general responsibility of the professsional CM is "to plan, administer, and control in a professional manner an overall construction program best suited to the individual project objectives of the owner" (Barrie and Paulson, 1976). It is recommended that he be involved at the beginning of the design process, along with the owner and architect, to preclude later architect–CM conflicts. The CM's duties may include the following: provide leadership to the construction team on all construction matters, inform the project management team of potential problems, provide recommendations of design or construction-method changes during the design as well as the construction phase, keep the owner informed of progress in terms of cost and time, and coordinate materials, equipment, and the work of all contractors (Barrie and Paulson, 1976).

Although the ASCE clearly identifies the construction manager as a professional, his true position is not yet clearly understood by the industry. If he is indeed a professional, then he acts as an agent of the owner, and thus his professionalism should ensure internal motivation, proper ethics, and reputable advice. If he is a vendor, then he is hired to perform his service at a price agreeable to the owner for a given level of quality. His work must then be supervised and checked to ensure proper performance. Generally, the former assumption is made, with the latter being attributed to general contractors who are not construction managers. As we see below, the project delivery system incorporates this distinction in its organization.

8.5.2 COORDINATION OF PROJECTS

The various phases of construction require communication and subsequent coordination between subcontractors and the general contractor, CM, or owner to ensure that the efforts are synchronized so that costly delays can be avoided. The coordination required during the project depends on the project's inherent uncertainty, as well as its size and complexity.

The type of coordination required in the project can be decided by analyzing the interdependence shared among the participants. Thompson (1967) has classified interdependence into three distinct categories (see Figure 8.10). Pooled interdependence is a relationship in which neither group has to interact directly with the other. In this case the coordination can be achieved by setting up routines, rules, or standards. Sequential interdependence implies that one group cannot begin work until the other has completed its tasks. Here coordination can be achieved by plans and schedules. In reciprocal interdependence, each group requires feedback from the other for mutual adjustments

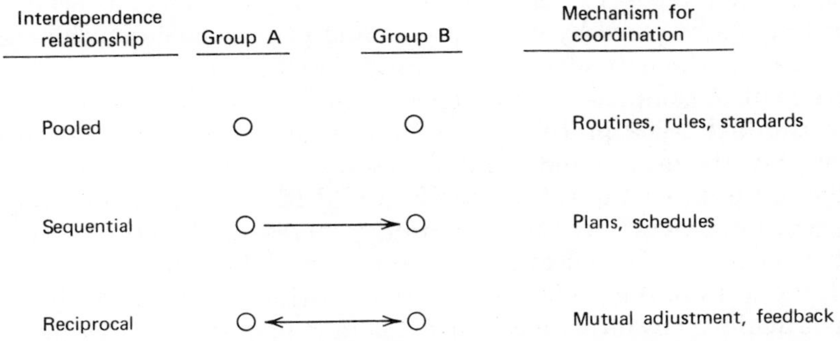

Interdependence relationship	Group A	Group B	Mechanism for coordination
Pooled	O	O	Routines, rules, standards
Sequential	O ⟶ O		Plans, schedules
Reciprocal	O ⟷ O		Mutual adjustment, feedback

Figure 8.10 Interdependence relationships (adapted from Thompson, 1972).

in the performance of their respective tasks. Hence pooled, sequential, and reciprocal interdependence require progressively increased coordination and communication.

The coordination problems are most obvious in large projects that require sharply differing tasks. For example, the construction of a pipeline in a relatively long but narrow project site leads to sequential interdependence. A power plant, on the other hand, which consists of diverse tasks performed in a concentrated area, results in a high degree of reciprocal interdependence. Modern, network-based analysis techniques, such as TREND (Benningson, 1972), attempt to provide an orderly investigation of the organizational requirements and interactions that exist.

8.5.3 PROJECT DELIVERY SYSTEMS

The owner has several project delivery systems from which to select his desired organization. Figure 8.11 depicts the traditional approach. The owner selects the architect–engineer and the general contractor. The general contractor in turn selects subcontractors to perform those special tasks that he cannot do with his own labor force. A variation of this approach includes a construction manager in addition to the architect–engineer and general contractor. The construction process is divided into the design and the build phases. As mentioned previously, the general contractor is typically not consulted during the design process, which might cause disagreements between the architect and the general contractor during the construction phase. The traditional project delivery system enables construction contracts to be bid competitively, after the phase design is complete. The need for owner involvement is relatively small. The designer coordinates his own work, and subsequently supervises construction. For this reason is has been, and continues to be, used by many public and private owners. The major drawback is the extended project duration which results from the requirement to complete design before commencing construction.

In the fast track process, the organization remains similar to the above, but the design and build phases are overlapped, resulting in a reduction in project duration (see Figure 8.12). Owing to the overlap, project interdependence and uncertainty increase. Thus there is a definite need for careful coordination of the work of the architect–engineer with the work of the prime construction contractors. This coordination can be achieved by either a construction manager or the owner himself. The design–build process integrates the design and construction into

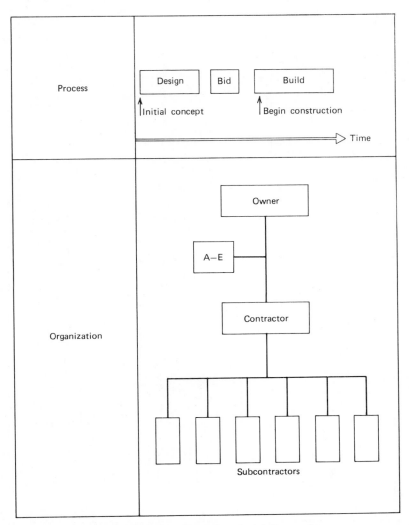

Figure 8.11   Traditional project delivery plan.

one single contract (see Figure 8.13). Here the owner specifies the performance requirements of the facility, and a single firm completes the design and construction with the design and building phases overlapping. The firm also manages the subcontractors. Because designer–constructor problems are resolved internally in this approach, the owner is less involved in the day-to-day decisions and consequently

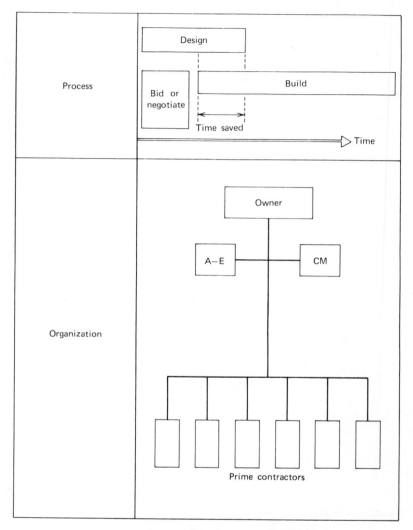

Figure 8.12 Fast track construction management.

has less need (and less ability) to exercise control over the evolution of the project. Again the duration has been shortened, cutting costs and also increasing coordination requirements. This procedure is especially effective in highly complex projects, with an owner who is not very knowledgeable about construction requirements.

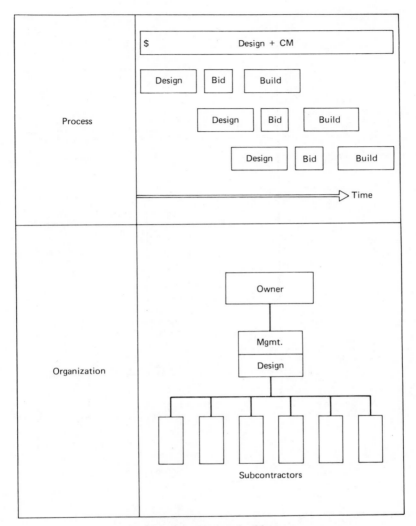

Figure 8.13 Design–build process.

In the turn-key approach, a single firm designs and constructs the structure, as in the design–build case, but it also finances the project (see Figure 8.14). The owner must ensure that his objectives are clearly understood by the design–construction firm early in the project definition stages.

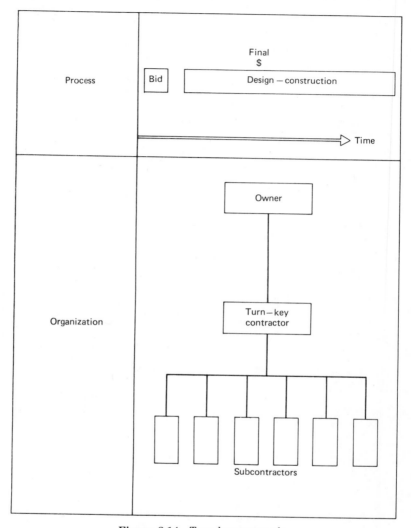

Figure 8.14   Turn-key approach.

8.6   THE STRUCTURE OF CONSTRUCTION PROJECTS

In the preceding sections we have examined the various types of organizations that are presently utilized by firms in the construction industry and the contractual relationships that may exist between project participants. We now describe the on-site organization of various types and sizes of construction projects.

The construction process has been described in some detail above. Briefly, it is initiated by the owner, assisted by an architect–engineer and other desired consultants, with a general contractor as project manager supervising contracted specialty firms. Successfully operating this collection of diverse organization requires skillful use of specialized management techniques.

8.6.1 SMALL AND MEDIUM SIZED PROJECTS

A typical project organization for building construction is shown in Figure 8.15. In this project, a supervisor coordinates the interaction

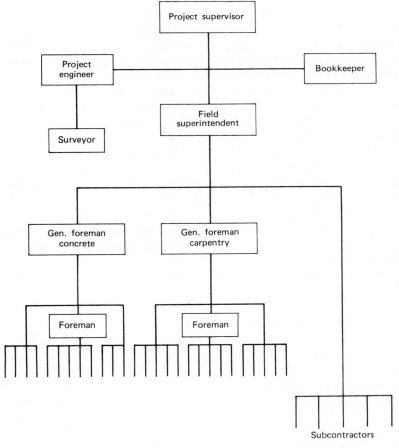

Figure 8.15 Typical small building project organization.

of the representatives of his own general contractor firm with those of a number of specialty subcontractors. The superintendent coordinates the entire on-site labor force, both his own and the subcontractors'. The general foreman at the lower level provides overall supervision for each of the different skilled trades. Usually the concrete and finish carpentry tasks are performed by the general contractor, whereas other skills are provided by specialty organizations that are subcontracted. This organization is a classical, functional hierarchy. Coordination between specialists is achieved through plans, schedules, and committee in weekly job meetings.

8.6.2 LARGE PROJECTS

In small or medium-size projects within the industry, the foreman has considerable authority in the performance of his assigned tasks. Larger-sized construction projects, such as nuclear power plants, are more complex and contain more levels of hierarchy. Their project organizations bear close resemblance to the characteristics of manufacturing industries. At the worker level performance is often specialized and frequently involves repetitive tasks. Since this type of construction is extremely complex with many interrelated components, the delegation of authority to the foreman is not possible. Decision making is usually limited to higher management levels such as area superintendents or engineers, whereas the function of the foreman is to supervise his crew so that production is maintained. The organization usually takes on a matrix form, where the general foreman is in charge of all crews in his trade but workers are also supervised in each work area by a technical coordinator to ensure that interdependent subtasks within each area are adequately coordinated (see Figure 8.16). Since tasks for workers on large projects are repetitive and relatively predictable, boredom often results. This is referred to again in Chapter 7.

8.6.3 DECISION MAKING ON PROJECTS

The unpredictable nature of the product and the environmental conditions that were emphasized in Section 8.4 typically necessitate decentralized management with a relatively autonomous project organization. Thompson (1967) has developed a table which describes the organizational task environment in terms of homogeneity, heterogeneity, and predictability–unpredictability (see Figure 8.17). The upper left quadrant, with a homogeneous and predictable environment, en-

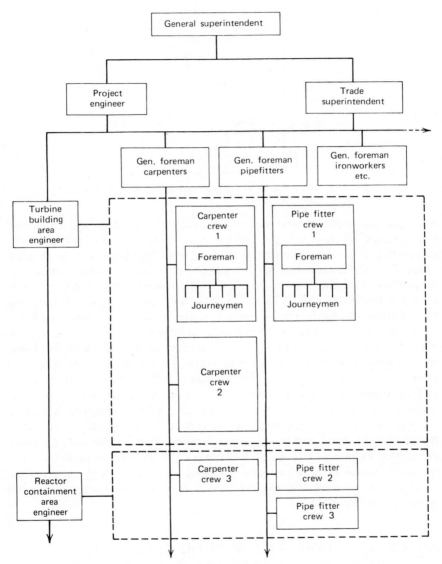

Figure 8.16 Matrix organization for large power-plant construction project.

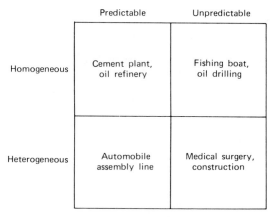

Figure 8.17   Dimensions of the organizational task environment.

courages the use of functional organizations with standard rules and regulations. In contrast, the construction industry operates in an organizational task environment that is heterogeneous and unpredictable, since the abundance of diverse activities creates a heterogeneous workplace, and the uncertainties that exist make the building procedures difficult to predict. This unpredictable nature requires frequent decisions at the project level which would be too costly and time-consuming if they were referred to higher levels within the organization. Thus the lower-level employees must be involved in the decision-making process to ensure a quick and accurate response to the frequent changes that develop in the environment of the construction task.

The degree of authority and responsibility assigned to each individual was described in Section 8.4. In the concluding section of this chapter we discuss how each participant can be given authority to fulfill his responsibilities and how his performance can be evaluated considering the extent of his responsibilities. The proper selection of performance measures evaluation such as time to complete the task, costs, and quality of work is important since it affects productivity as well as work satisfaction.

8.7   ORGANIZATION OF WORK AT THE CREW LEVEL

The organization of construction firms and projects have evolved in response to the product, the technology used to create it, and the environment of the construction process. The work crew is the most important element of a construction project, and its organization has many unique traits that have developed in a similar manner to the project organizations described above.

8.7.1   CREW STRUCTURE

The Work Crew

The work crew can be considered to be a mini-organization of its own. It is usually a well-defined group whose members may vary over time, but is well-known at any one time. The crew members are motivated to work both by the goals of the project and by the specific human needs of each individual. The work activities of the crew are usually clearly defined and, in the case of union crews, are explicitly stated in the form of jurisdictional specifications. Just like the mother organization the crew functions in an environment of uncertainty, and reacts to this uncertainty so that it can anticipate and thereby reduce its impact.

One common response of an organization to uncertain work conditions is to establish a clearly defined boundary for its activities. The boundary can then be changed regularly to meet the needs of the changing environment. A construction work crew's size and composition are determined by the work to be done, and can be changed daily if necessary. Definition of the area of work partly depends on the crew itself; members are chosen based on their training, experience, and skill in specific trades. Much effort is made in construction companies to agree on trade jurisdiction so that a crew's responsibility and authority are clearly known. (The relative location of authority is a most important concept in understanding the construction industry; we return to it shortly.) According to Thompson's (1972) model the individual workers in a construction crew interact with each other within the boundary of the group applying their technology or craft to get the job done. Various boundary-spanning elements, such as contacts by the foreman, can transfer information into and out of the crew organization and protect the ongoing work from the influence of outside uncertainty. The foreman receives goals from project management, plans the work, settles conflicts with other crews, and ensures adequate and timely availability of materials and plans. Subcontracts, labor agreements, and jurisdictional specifications, which are explicitly or implicitly agreed to by the crew, also reduce uncertainty for the crew.

Union Work Crew

There are several differences between union and nonunion work crews. In the union situation, the crew has a foreman and a number of journeymen and apprentices. The majority are journeymen, most of whom are trained in skills appropriate to their trade. They are to a large degree self-sufficient, and require relatively little direct supervision. An apprentice may perform some of the unskilled or semiskilled labor that does not require the capabilities of a journeyman. An apprentice is a trainee who is enrolled in a formal training program with the goal of becoming a journeyman. Other avenues to the status of journeyman are available, however. The union may grant journeyman status to informally trained—or untrained—workers, certifying that they possess the required skills and experience to practice the trade. Demand for labor and competition from the nonunion sector can have a strong effect on the use of such measures. The maximum ratio of apprentices to journeymen is limited by labor agreements to about 1:5 for most trades, but varies with the situation and the trade, as does the proportion of journeymen having apprentice training. In the mechanical trades (pipefitting, plumbing, and electrical work) where state licensing is required, more than 80% of journeymen have had formal apprenticeships. On the other hand, the proportion of apprentice trained carpenters is around 30%, with laborers and teamsters lower still.

The journeymen and apprentices are led by the foreman, who (ideally) is very experienced and capable of skilled work in all areas of his trade's jurisdiction. He coordinates the workers' efforts, makes decisions referred to him by the journeymen, and spans the boundary of his work crew in a number of ways. He is the link between labor and management; his authority may vary from being similar to other journeymen on large projects to being the subcontractor's senior representative on small projects, with complete responsibility for the completion of the contract.

In some cases, union contractors have little say in the selection of foremen. This obviously reduces management's ability to use selection of foremen as an incentive. Furthermore, foremen usually receive only a very small wage increment relative to journeymen, typically $0.50 to $1/hr. Hence there may be little incentive for a union worker to become a foreman. Secondly, a worker in a foreman's crew one day may be his foreman the next. Consequently his authority, and his loyalty to management, must be tempered by this consideration. The above problems apply especially when out-of-town contractors try to assemble crews quickly from the local labor unions' currently unemployed membership, or in times of peak construction activity.

Nonunion Work Crew

In a nonunion crew, the division of skill and responsibility levels is not nearly as distinct as for union crews. A continuous spectrum of abilities exists. This is reflected in wage data presented by Bourdon and Levitt (1980). Wage distributions for union and nonunion workers in the Boulder–Denver area are shown in Figure 8.18. The average nonunion wage level is lower than that of the union journeyman; however, some nonunion workers earn even more than a union foreman. The variation in skill levels in nonunion crews requires more direct supervision, as evidenced by the leadmen who assist the foreman in this area of his responsibilities. A trade-off exists between lower wage scales and increased need for supervision. Less restricted trade jurisdictions lead to less specialization in nonunion labor.

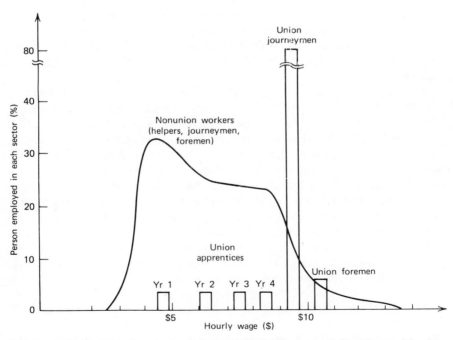

Figure 8.18 Union and nonunion relative wage levels (SOURCE: Bourdon and Levitt, 1980).

Formal and Informal Groups

The construction worker is a member of both formal work groups, such as his crew, and informal groups, such as those he eats lunch with on the job or spends leisure time with off the job. In the construction industry, these groups tend to be more closely related than in other industries. The nature of work methods causes more personal interaction than might be found in an industrial production situation. The solving of problems at the worker level strengthens their interdependence. This closeness between formal and informal groups reinforces group unity and makes construction crews very cohesive. Used carefully, this can positively effect production and quality of work. In one case, production was almost doubled when healthy competition was encouraged between crews on twin towers of a high-rise building.

The formal organization attempts to motivate worker behavior, through pay and other benefits, to maintain a desired level of productivity and quality, which will result in an acceptable product that is completed on schedule. Yet the worker also may be pressured by his peers to work at an established pace that will not embarrass his slower co-workers. An informal, established work rate may be enforced through peer acceptance in work groups and recreational activities, such as bowling teams or drinking groups. Thus the strength of the informal organization in the influencing of worker behavior cannot be ignored.

Management must recognize the existence of the informal organization and its goals, as well as the formal structure. Rather than oppose the standards and rewards of the informal group, management must attempt to influence informal group goals in the same direction as the goals of the formal organization. Then the worker can simultaneously satisfy the formal demands of the firm as well as informal demands of other groups. To achieve this it can be recommended that management and labor make a unified effort to develop a team spirit, for instance through the increase of personal interest in the worker's welfare, short victory celebrations when major milestones are met, and other similar measures. Such actions may be considered quite elementary and even parochial but the basic intent is to develop a harmonious team dedicated to a common goal.

Effect of Worker Specialization

The efficiency of a crew on a given project is partly determined by the occupational breadth and skill levels of its members with respect to the size of the project. Differences in breadth and skill level between union and nonunion workers have implications where project size is considered. Bourdon and Levitt (1980) conclude that the relative efficiency of union versus nonunion labor may be primarily related to the appropriateness of the unions' occupational breadth for a given type of construction. The best indicator of this seems to be the size of the (peak) on-site labor force on a construction project. As explained in Section 8.5, the unions tend to dominate the medium-sized projects, whereas nonunion firms are strong in very small- and very large-scale construction.

Uncertainty and Communication at the Crew Level

Galbraith (1975) has shown that as organizations work in the same area, uncertainty in the work will require them to refer some decisions to higher authority. When interdependence exists between the functions, coordination between them is needed. For individual workmen, the foreman can do this; for crews in different trades or physical locations, coordination between foremen or through a third party such as a general foreman takes place. Figure 8.19 shows the situation. Each of two crews has a foreman and a number of workmen. The environment contains elements of uncertainty that affect each crew as it carries out its work. Workers can refer decisions or needs to the foreman for resolution. The foremen can coordinate directly with one another or through the project management. Each of these transactions requires a channel of communication, time to be accomplished, and an amount of effort on the part of the people involved. Communication of status and decisions is critical at the crew level within the organization. Workers must be able to interact with their fellow members, as well as recieve orders from the foreman, and provide feedback to higher levels.

The crew is also linked to an informal communication network, or "grapevine." This is a powerful communication channel which is frequently faster, although often not as accurate, as the formal one. Management can monitor the grapevine to ensure that no damaging information is passed. Frequently the informal channel provides prior warning of problem areas, such as worker dissatisfaction, that is not provided by formal means. Thus management can find a more candid indication of the project status that may be distorted in formal reports. It may also utilize the formal communications network to dispel false rumors generated in the grapevine before serious results have developed.

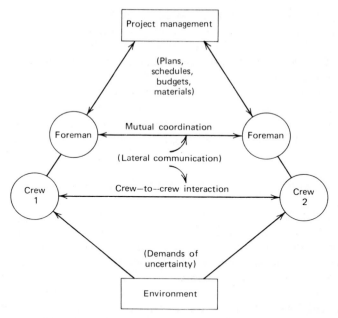

Figure 8.19 Uncertainty and communication at the crew level.

Decentralization

As the work environment becomes more uncertain (unpredictable) more exceptions to plans occur, requiring more communication between crews. As this increases beyond a certain point, the hierarchical information channels become overloaded (Galbraith, 1973) and workers and foremen must resolve problems through lateral communication. Decentralization has occurred *de facto* (see Figure 8.19).

In the construction industry, the professional training of manual workers and the decentralization of decision making down to the work level provides an efficient and economical solution to production administration. The foreman and often the workmen themselves perform the boundary-spanning function by transmitting needed information and assuming responsibility for decisions made. The mass production industry provides an "industrial engineering" department that delineates the method of work at the bottom level. In constrast, the construction industry normally gives only product specifications and cost targets to the workers, in the form of budgets, blueprints, and local regulations. The method of producing the desired product is not designated. Thus the foreman in construction can utilize his own procedures for production within the economical, regulatory (building codes, etc.), and technical constraints of his craft. Naturally, the delegation of decision-making responsibility places additional emphasis on the competence of the foreman and journeyman in the construction industry. However, each level in an organization has limits of authority and responsibility. These limits are a response of the organization to the degree of uncertainty in its environment (see Preceding section) and have a large impact on the effectiveness and efficiency of the organization. The authority and responsibility assumed by an individual are important determinants of his motivation. We now consider motivation in detail.

8.7.2 MOTIVATION OF THE WORK CREW

The organization of work in the construction industry directly affects the forces of motivation (and satisfaction) that are important for the workers. A review of behavioral theories leads to an understanding of motivation and small group dynamics; we shall see that management can adopt a unique perspective toward motivation in construction; we then go on to examine the impact of the union structure on motivation and special problems associated with very large construction projects.

Theories of Motivation

The work of Maslow (1954) is familiar to any student of motivational theory (see also Chapter 7). The basic tenet of this theory is the existence of five types of individual needs, arranged in a pyramid: physiological, safety, affiliation, self esteem/ego, and self-actualization needs. An individual is motivated to satisfy these needs, beginning at the bottom level and working his way up to the next level of the pyramid only when a given need is largely satisfied. At any time, the lowest need that is not satisfied will be the dominant motivation. The importance of each need to the individual is subject to change over time or in different situations. The first two needs can be satiated (i.e., they are finite), whereas the upper three levels do not have a satisfaction limit. In well-developed nations, the importance of the physiological and safety/security needs has been of decreasing significance with the increase in the general standard of living. Hence the higher type of needs are now considered to be the critical determinants for individual behavior. This certainty also holds true for employees of the construction industry.

Herzberg (1966) further divided Maslow's needs into two categories: hygiene factors (equivalent to physiological and safety/security needs) and motivators (affection, self-esteem, and self-actualization). Whereas Maslow was more concerned with the individual, Herzberg focused his attention on the environment. Herzberg's two-factor theory, as it is called, includes recognition, advancement, responsibility, growth, and achievement as motivators (Luthans and Ottemann, 1973). He equates the hygiene needs (e.g., salary, working conditions, company policy) to *job dissatisfiers* and the motivators to *job satisfiers*.

Maslow's and Herzberg's theories have been criticized and modified in subsequent investigations of behavioral scientists (see Chapter 7).

Motivation of Construction Workers

The motivation and work satisfaction of a construction worker has several sources. The individual can assume authority and responsibility with respect to the project, exercise skills and abilities in doing the work, and interact with the crew in a close manner. The group of workers around him can provide support, companionship, affection, and establish a basis for his self-esteem. In addition the product of construction work is permanent and often monumental in size, which can strongly reinforce the worker's sense of self-actualization.

Work as a Motivator in Construction

Borcherding (1972) pointed out that in construction the basic needs (as presented by Maslow) are not as dominant as the higher needs. The basic needs are already primarily satisfied, and thus have little effect on worker motivation. Rather, the construction industry provides the worker with an occupation that results in a physical "monument" to his efforts. He has a well-defined goal: the structure itself, which when completed instills a sense of accomplishment. Alienation of the worker from the overall task, as is prevalent in mass production, is thus rare in construction. Only the very large projects contain repetitious specialization and thereby eliminate the participative decisin making that is a characteristic trait of the construction industry. The construction worker can therefore achieve considerable satisfaction of his needs for self-esteem and self-actualization that normally are not available to fellow workers in manufacturing industries.

Construction workers are frequently laid off after the completion of a project. Union organizations provide hiring halls to simplify the acquisition of construction workers by management organizations. They also provide negotiated wages, and benefits that help the union worker to survive in periods of unemployment. The less skilled nonunion worker does not receive these union advantages. Thus job security is a more critical factor to nonunion construction workers. Yet nonunion as well as union construction workers accept this characteristic of the industry, and thus their motivation for security, which is not guaranteed by the organization, is typically not a crucial factor in the determination of his behavior.

Satisfiers and Dissatisfiers for Construction Workers

Borcherding (1972) carried out a survey of construction workers' job satisfiers and dissatisfiers. He found that the foreman perceives the challenge of running the work, the maintenance of the job schedule, and good workmanship as his primary motivators. His dissatisfiers include uncooperative, unmotivated workmen and lack of proper management support. The journeyman and apprentice find that completed tasks, a tangible physical structure, and social work relations are their most important satisfiers. Their dissatisfiers consist of poor interpersonal relations, unproductive workmen in their crew, and poor workmanship. Borcherding offers recommendations for improving satisfaction and reducing dissatisfaction. Primarily they consist of the practice of good management techniques by superintendents to keep the work moving.

This is a crucial difference between construction and other mass production industries, in which supervisors at all levels must be very concerned with motivating their subordinates. The literature of organizational behavior is filled with articles suggesting job enrichment, participative decision making, and other devices to restore motivation to alienated workers in mass production or clerical situations.

Construction work, on the other hand, is inherently rich in these "devices" of motivation. The work is seldom repetitive; workers are actively involved in decision making and the crew organization described above leads to a rich array of social interaction between crew members. Management is thus free to plan, schedule, and allocate resources—motivation will take care of itself if they do this well!

Effect of Competition and Collaboration

Competition is a pervasive characteristic of U.S. culture. It has been employed with some success in the construction industry in order to increase worker motivation and hence productivity. Crews compete with each other in the construction of tunnels, floors of office buildings, and other such repetitive, group tasks. The major disadvantages of this technique are the increase in safety risks and the potential decline of quality standards. Management must be aware of these problems and establish procedures that encourage their elimination. For example, penalties can be assessed for safety violations or substandard performance. Furthermore, the establishment of union regulations precludes the use of some kinds of competitive techniques.

Another disadvantage is that the competitive spirit might lead to a decrease in collaboration between crews. This decreases information flow and hinders problem-solving procedures. Thus the obstruction of information flow must be outweighed by the benefits of a competitive spirit if competition is to be worthwhile. When the competitive spirit is not generated, workers and foremen may assist each other, in a procedure that has been characterized as an exchange of knowledge for an indication of superiority. When competition between crews is present, this assistance is still encouraged, but only within the limits of individual crews.

Management Impacts on Productivity

Thus motivation is not a critical factor to construction management in the way it is in other industries. Logcher and Collins (1978) point out that this situation provides a unique opportunity for construction management to concentrate its efforts on organizational and management tasks rather than on worker motivation and company morale. They measured a four to one difference in the rate of floor tile laying attributable to variations in management alone. Clearly, in most construction situations, the highest return for management efforts is derived from increased emphasis on planning, scheduling, and organizing, rather than from attempting to improve motivation directly.

Impact of Union Structure on Motivation

Unionization can effect the motivation of workers by reducing their desire for promotion. The foreman's position as a focal point of communication and decision making between his crew and both the union and project management can lead to conflicts any time the union and management are in opposition. The foreman's job is therefore often stressful and a journeyman may choose not to improve his skill level nor assume more responsibility in order to avoid a promotion to this position of conflict.

Gradual differentiation of skill levels in the nonunion sector allows graduated changes in pay and status to be made to reflect a worker's effort and ability, thus providing a continuous incentive for advancement.

Hence the union structure clearly has disadvantages from a motivational point of view. However, the fixed and uniform wage scale provides contractors with an important benefit; it enables them to estimate labor costs with a higher degree of confidence, and to be confident that they can assemble a work force with known skill levels at the estimated hourly wage.

From a contractor's point of view, therefore, a trade-off exists between loss of motivation and the degree of confidence he can have in a project's labor supply and estimated cost.

Special Problems of Large Projects

Increased size brings problems in organizational structure and has impact on the motivation of workers. A large size promotes a high degree of worker specialization, which requires more coordination and communication in both the vertical and lateral dimensions of the organization. Centralized decision making tends to result, and the worker may have a sense of remoteness from the finished product. Often disheartening rework is caused by delayed changes in specifications and slow communication. This has negative effects on work motivation and brings about less self-actualization, and more alienation and boredom. A higher turnover rate and absenteeism result, which defeat the productivity benefits from the increased speed of learning the more specialized work tasks.

Techniques for Developing Motivation on Large Projects

Some techniques have been proposed to combat the effects of project size on worker satisfaction and productivity. This type of construction work has several similarities to work in the manufacturing industry. Borcherding (1976) describes two areas of management actions available to encourage worker motivation in the construction of large and very large projects.

Organizational Changes

Borcherding suggests that one should avoid deep heirarchical structures by decentralization of responsibility and authority as much as possible, and by establishing as few levels in the hierarchy as is required to maintain coordination and control.

One way to decentralize is to subcontract more of the work to local contractors who have established organizations and work methods. Division of the work into geographic areas of responsibility or division by work specialization also helps to ameliorate the motivational problems of large projects.

However, large, complex projects, whose subactivities are highly interdependent, do not permit decentralization without a loss of coordination between the subtasks. This is the problem with power plants and refineries, where Borcherding gathered his data.

Motivation of Workmen

Although management attention to motivation has been said to be unimportant for most projects, we have described in the preceding section how it can be significant on large projects. This section lays out some suggestions for attacking motivational problems, particularly on large projects. A number of specific steps can be taken that increase the work satisfaction and work motivation of the individual worker.

1. If supervision of a worker is reduced while his accountability for the results of his work is maintained or increased, he feels an increased degree of responsibility.

2. Giving a person an individual unit of work increases his satisfaction in its completion.

3. Granting additional authority (where deserved) improves a worker's self-esteem and provides a motivating example for others.

4. Making progress reports available to the worker himself increases his identification with the work.

5. Assignment of new or harder tasks in proportion to ability provides a challenge and establishes a norm of increasing skill levels.

6. Giving tasks requiring special skills consistently to a particular worker allows him to develop a degree of specialization and expertise, gaining him the respect of his peers and contributing increased value to the organization.

7. Encouragement of worker initiative in performance of the work, and especially in the preparation phase of the work, promotes the development of more efficient methods.

8. Recognition of innovation and a tolerant attitude toward trial procedures encourages positive contributions and increased productivity.

Motivation of craftsmen can be indirectly encouraged in a variety of ways:

9. Short orientation sessions for new employees give workers an early sense of identification with the project.

10. A monthly project newspaper is an excellent communication tool that keeps workers informed of other aspects of the project and lends a sense of purpose to their daily work.

11. Biweekly luncheons with selected crews can add to their sense of unity, call attention to above-average performance or notable safety records, and acknowledge a particularly difficult period or phase of the construction.

12. Sports teams, on-the-job dinners, and training courses encourage the workers' sense of unity and companionship.

13. Tours of the site for workers' families and members of the community give meaning to the project and establish the workers as active members of the society who are making a positive contribution.

Worker motivation is an important aspect of the success of construction projects, both in the resultant quality and in the maintenance of scheduled completion targets. The many techniques described above are most applicable in large projects, but have application in all areas of construction. Yet it has been established that good organization and management are the most important factors for increasing worker motivation and productivity (Logcher and Collins, 1978). Section 8.8 explores some of the measures of performance used in the industry and constraints on their use.

8.8 MEASURES OF PERFORMANCE IN CONSTRUCTION

We have investigated the organization of work in the construction industry at all levels. In this concluding section we examine some measures of performance that are used within the industry both to evaluate its employees and to monitor the progress of current projects.

8.8.1 MULTIPLE GOALS FOR ORGANIZATIONS

An organization provides measures of performance for several types of functions. The evaluation of the overall performance of a company must include a consideration of long-term goals as well as short-term. Long-term goals such as customer satisfaction and quality control become important considerations in maintaining a favorable reputation and ensuring future negotiated contracts within the industry. In addition, decreasing labor turnover, maintaining acceptable levels of safety, and continuing good labor-management relations are important factors in the successful operation of the firm. Simon (1976) has characterized these multiple goals as a series of constraints that the organization must face in the attempt to attain its primary objectives, such as growth and profit. He has equated this situation to the basic linear programming problem, in which the objective function (a single goal) is optimized subject to certain specified constraints (all other goals). The consideration of multiple goals in the operation of an organization increases the difficulty of achieving an optimum level in each component since there might be complex interdependencies between the different goals. Simon suggests that when confronted by multiple goals, organizational participants attempt to "satisfice" rather than maximize them. This means that a performance target is developed for each goal, and only when it is met at a satisfactory level is the next target considered; and so on. The advantage of this procedure is that the multiobjective maximizing problem is made into a series of simple checks of the form, "Is performance better than or equal to the target for this goal?" (see Figure 8.20).

However, the procedure does have some disadvantages. Specifically, the cyclical nature of attention to goals can result in a continuous oscillation of performance between two targets (e.g., costs and accident rates) which cannot both be satisfied. In this case, targets must be reevaluated and adjusted to an attainable level. This can be equated to a "squeaky-wheel" theory of the management of the organization. Simply stated, the "squeaky-wheel" theory means that the problem (goal) that makes the most noise (i.e., is most serious) receives attention first, at the expense of all others. Thus the worst performance level is increased until it achieves the desired target level. Then the next most serious problem is considered.

8.8.2 MULTIPLE GOALS IN CONSTRUCTION

The characteristics of the construction industry, including the temporary organization of the project and the intensive use of labor, emphasize the measures of production rate and unit cost in the measurement of company success. Schedule consideration is also frequently transformed into units of cost. Thus the single consideration of expended cost generally becomes the paramount goal and receives the

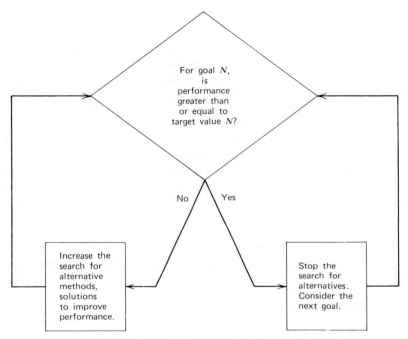

Figure 8.20 "Satisficing" with multiple targets.

full attention of project managers and foreman, often at the expense of other important factors such as quality and safety. The use of one such measure simplifies the evaluation process, yet it is important to provide a clear relationship through the use of appropriate trade-off values between the short-term, direct-cost performance measures and other longer-term cost considerations, such as safety and quality, to ensure that the latter are not ignored. A more appropriate strategy that should be employed by management is to develop performance variables that can adequately take into consideration quality and safety as well as time and costs. Presently many construction organizations fail in this respect because they utilize their established *financial accounting systems* for the management of their projects. The accounting system, although it adequately monitors the financial transactions of the company, does not sufficiently consider such factors as safety and quality control. A separate *management accounting system* is therefore required. Such accounting systems are feasible and inexpensive today using modern computer technology.

Safety in Construction

Insurance costs and other accident expenses are a significant factor in the overall profit of the firm. Yet many company presidents report that their project managers do not pay enough attention to safety. Levitt (1975) conducted a survey of 23 construction firms in order to determine the reasons why some companies had very high accident costs and others had low accident costs. He found that when top management both knew and acted on information about accident rates, the accident costs were reduced to 25% of those of companies whose top management either did not know or did not use safety information (Levitt, 1975).

Thus management can encourage the avoidance of accidents and the use of safety equipment by employing measures of performance that include safety costs and benefits. The long-term goals associated with an acceptable level of accidents, such as reduced insurance costs, can be transformed into short-term goals for the lower levels of the organization by incorporating appropriate costs in the cost reporting system.

Quality Control in Construction

Quality control is seldom used in formal performance considerations. However, it is important that a high level of quality be maintained to ensure customer satisfaction and future business. Yet frequently quality is sacrificed for an increase in productivity and a decrease in direct costs. For example, a company's construction of pipelines utilized competition among crews that completed different sections of the project. Although this encouraged productivity, the subsequent cost of repairing substandard welds significantly increased the final cost of the project. This example illustrates the need to use quality performance measurements in addition to other performance measurements. Quality considerations may be introduced by assessing the costs of rework and informing the crews involved and by charging all project rework to the project's cost. This also motivates workers to adopt the long-term goals of the organization which are otherwise not made explicit by the cost reporting system to which the foreman has access.

8.8.3   BEHAVIORAL ASPECTS OF CONTROL SYSTEMS

Levitt and Logcher (1976) investigated the human element in project control systems, and pointed out that as projects or companies become large, formal control systems become increasingly important in the evaluation of individual performance. The control variables typically used in construction consist of time, production costs, safety costs, and quality, which are employed in various combinations at different levels within the project or company organization. The individual employee recognizes these control variables as performance measures and usually attempts to maximize his performance in these terms. Selection of proper control variables is hence important since otherwise the individual will not be maximizing the performance of the overall organization. One such example is the consideration of safety as described above (costs for actual accidents plus costs for accident prevention). Let us consider it again, in terms of the effect of project control systems on safety.

Frequently a project is charged only for accident prevention costs (posters, safety nets, etc.) and not for the total safety costs, which also include costs for company insurance. A project manager might therefore choose to minimize the accident prevention costs. Although this action temporarily increases the "profitability" of the project, it is likely that accidents will increase. This in turn increases the company's "experience modification rate" (this is a safety statistic calculated by state Insurance Rating Bureaus and is used by the insurance companies to calculate premiums), with the result that future insurance costs also increase. Levitt (1975) showed that a simple cost accounting device—"charging" insured accident costs to projects immediately—produced enormous improvements in project accident rates.

The allocation of control variables to project participants must include a consideration of authority and responsibility. Clearly the individual must be able to control those areas in which he is being evaluated. The project manager is generally evaluated in terms of cost and time (i.e., maintaining scheduled target dates). Quality may also be considered, but frequently both quality and time are combined into some form of penalty costs. As discussed previously, safety may also be included in this performance measure. The single variable reduces the difficulty in objectively measuring performance, and is accurate only when the trade-off ratios that are used are carefully determined.

The foreman is evaluated in terms of productivity, which is associated with both cost and time. In addition, quality considerations and safety records may also be incorporated in his performance measure as done in the case of the project manager above. As previously discussed, crew unit cost reports are the primary means of evaluating the performance of the foreman (Borcherding, 1977). Subjective evaluations of safety and quality are secondary considerations.

Thus cost is the most significant performance measure used for the evaluation of project managers and foremen. Nonmonetary criteria, such as safety and quality, outweigh cost considerations only when a serious crisis—a "squeaky wheel"—develops in either area. A record of bad accidents shifts managerial emphasis to safety, whereas frequent reports of customer dissatisfaction encourage a greater demand for quality performance. But once this crisis has passed, the evaluation criteria swings back to the cost consideration. Thus a series of oscillations results.

The utilization of a single cost control variable, which effectively incorporates safety and quality concerns at each level, discourages this oscillation from one problem area to another, and results in consistently balanced performance. This single control variable could just be *a more complete and accurate measure of all long- and short-term production costs*. For example,

C = direct labor costs + direct material costs, + indirect labor costs + other indirect costs + true accident costs + discounted present value of future costs due to rework + . . .

However, this requires a change in philosophy and in control procedures for construction firms, and will probably be slow to evolve in the industry.

8.9 RECAPITULATION

In this chapter we started with an analysis of the construction product and its influence on the characteristics of the industry. The unique nature of the product, as well as its uncertain environment, determines the characteristics of the industry: labor-intensive, flexible, fragmented, and local. Trade specialization and decentralized decision making are necessary to deal with the unpredictability that commonly exists at the job site.

The construction process requires extensive personal communication, particularly for employees who perform special trades or professional tasks. The complexity of present-day construction projects requires the skillful management of interacting firms to ensure proper coordination. The project delivery system that is employed is a function of project complexity, time considerations, management capability, and the degree of owner involvement.

The organization of firm is influenced mainly by two factors: *the requirements dictated by its projects and the technology of the firm's tasks*. The selection of functional, matrix, or project structures depends on the required level of coordination, the capabilities of the firm's project managers and department chiefs, and the characteristics of the project (e.g., technology, duration, size, interdependence). The types of firms present in the construction industry, that is, architect–engineers and construction contractors, utilize matrix or project organizational structures. The special characteristics of the owner-agency firm, such as an in-house construction organization, were also discussed to highlight the trade-off between fast decision making and owner control of the project.

After examining the construction firm, we then focused on the organization of construction projects. The unpredictable nature of the product requires that decision making be delegated to the foremen and journeymen. The lack of participative decision making in large industrial construction projects is a result of their increased complexity and hierarchy, these result in a tremendous amount of coordination being required owing to the uncertainty that is present and the multiple interdependence that exists between the components of such projects.

The organization of work at the crew level was then examined. the journeyman is supervised by a foreman. Both individuals attempt to maximize their performance in terms of measures that are established by the organization. The skill level and supervisory requirements differ between union and nonunion workers. The decision-making authority of the foreman is unique to the construction industry.

The influence of small group dynamics was found to be strong in the construction industry. The motivators that are associated with journeymen and foremen include the tangible nature of the product and the existence of participative decision making in the construction process. The informal organization is especially important in such a labor-intensive, floating-work-force industry. Competition, within bounds, can be successfully employed to encourage increased productivity. Communication within the group is especially important in such a close-working craft industry.

Finally, we discussed the measures of performance that are used to evaluate individual efforts, and thereby to motivate behavior. Desired performance targets are frequently "satisficed" rather than maximized. The temporary organization of projects and the intensive use of labor encourage productivity measures in terms of cost. The danger of excluding safety and quality considerations until they have reached a crisis level of poor performance is present in many management systems currently employed by firms in the industry. The combination of long-term goals and short-term in the industry. The combination of long-term goals and short-term monetary goals into a single measure is necessary to ensure balanced attention to all important factors within a construction company.

REFERENCES

Barrie, D. S. and Paulson, B. C. Professional Construction Management. *Journal of the Construction Division ASCE*, **102**(CO3), September 1976.

Benningson, L. A. TREND: A Project Management Tool. *Proceedings of the INTERNET Conference*. Stockholm: 1972.

Borcherding, J. D. *An Exploratory Study of Attitudes that Affect Human Resources in Building and Industrial Construction* (Stanford Univ. Tech. Rep. No. 159). June 1972.

Borcherding, J. D. Applying Behavioral Research Findings on Construction Projects. *Project Management Quarterly*. **7**(3), 9–14, September 1976.

Borcherding, J. D. Improving Productivity in Industrial Construction by Effective Management of Human Resources. *Project Management Quarterly*, **7**(3), September 1976.

Borcherding, J. D. Participative Decision Making in Construction. *Journal of the Construction Division ASCE*, **103**(CO4), December 1977.

Bourdon, C. C. and Levitt, R. E. *Union and Open-Shop Construction: Compensation, Work Practices and Labor Markets*. Lexington, Mass.: D.C. Heath & Co., 1980.

Building & Construction Trades Department, AFL-CIO. *Agreements and Decisions Rendered Affecting the Building Industry*. Washington, D.C.: 1973.

Galbraith, J. *Designing Complex Organizations*. Reading, Mass.: Addison-Wesley, 1973.

Herzberg, F. *Work and the Nature of Man*. Cleveland: World Publishing, 1966.

Levitt, R. E. *The Effect of Top Management on Safety in Construction* (Stanford Univ. Tech. Rep. No. 196). July 1975.

Levitt, R. E. and Logcher, R. D. The Human Element in Project Control Systems. In *Project Management Institute Annual Symposium*. Montreal: October 1976.

Logcher, R. D. and Collins, W. W. Management Impacts on Labor Productivity. *Journal of the Construction Division ASCE*, March 1978.

Logcher, R. D. and Levitt, R. E. Principles for an Integrated MIS for Design Firms. *Second International Symposium on Organization and Management of Construction of CIB W-65*. Haifa, Israel: October 31–November 2, 1978.

Luthans, F. and Ottemann, R. "Motivation vs. Learning Approaches to Organizational Behavior." *Business Horizons*. December 1973.

Maslow, A. *Motivation and Personality*. New York: Harper, 1954.

Mills, D. Q. *Industrial Relations and Manpower in Construction*. Cambridge, Mass.: M.I.T. Press, 1972.

Northrup, H. R. and Foster, H. G. *Open Shop Construction*. Philadelphia: University of Pennsylvania Press, 1975.

Simon, H. A. *Administrative Behavior*. New York: The Free Press, 1976.

Thomsen, C. B. *How to Buy Design and Construction*. Houston: C.M. Associates Inc., undated.

Thompson, J. D. *Organizations in Action*. New York: McGraw-Hill, 1972.

U.S. Department of Commerce, Bureau of the Census. *Census of the Construction Industry*, 1972.

U.S. Savings and Loan League. *Savings and Loan Fact Book—1970*. Chicago: 1969.

Youker, R. Organizational Alternatives for Project Management. *8th Annual Symposium, Project Management Institute*. Montreal: October 6–8, 1976.

CHAPTER 9

CONSTRUCTION PRODUCTIVITY

DONALD C. TAYLOR, Ph.D.

Construction Industry Research
Denver, Colorado

JACK W. WARD

Arizona State University
Tempe, Arizona

ROBERT F. BAKER

Engineer and Policy Consultant
Bethesda, Maryland

The purpose of this chapter is to clarify the concept of productivity and related issues in the construction industry. Because construction is such a vast, fragmented, and complex industry, it is often as hard to identify and define the problems of productivity and cost effectiveness as it is to resolve them. There are many reasons, however, to try to understand productivity and the problems involved in improving it, even though there is no universally applicable solution.

There is considerable evidence (Taylor et al., 1976; Thomas et al., 1971) that improved productivity in any industry is related to levels of research and development. The current research and development effort in the construction industry is very small compared to even moderately progressive industries in the United States. Agriculture, which is highly R and D oriented, claims an annual rate of growth in productivity in the order of 8%. The productivity growth rate for construction is in the range of 1 to 2% (Cassimatis, 1969).

Increased productivity and increases in construction activity in the past have been largely due to the development of more efficient materials, tools, and machines. Most of these developments resulted from research and development efforts of the manufacturing industry. For example, earthmoving equipment has continually improved over the years. Larger, faster, more powerful, and more efficient machines mean that one operator moves increasingly more earth per hour. But since wages also rose, the cost of earthmoving stayed about the same and the potential productivity gains were offset. Technological advances in tools and machinery were less dramatic after the 1950s and, as a consequence, construction unit costs have increased.

The output of construction workers is often increased by the introduction of labor-saving tools: for example, a power saw in place of a handsaw or a power trowel instead of a hand trowel increases the output of an individual worker in a given period of time. But the increases in output that derive from using power tools and equipment depend on increased energy use (electricity, gasoline, or diesel fuel). Thus trade-offs exist, and increased job-site production that requires increased energy costs may not necessarily increase total productivity for the project (or for the broader economic or social system). It becomes clear that the context, terminology, and measurement of productivity are critically important in discussing or comparing actual productivity.

Generally, people have a number of critical misconceptions about the mechanisms of the construction industry. These lead to costly errors in the design and construction of new facilities. The decision-making and policy-development processes that must occur prior to and during construction often partake of these misunderstandings. This, in particular, makes cost-effective planning very difficult. The following list contains some of the typical misconceptions that make increased productivity hard to achieve.

1. The industry is craft-oriented and must build in this mode.

2. The industry is so fragmented and diverse that nothing can be done to alter the product-delivery system.

3. Each facility must be unique.

4. There is little that can be done to improve unit productivity.

5. There is little that can be done about the liability or risk issue that is stifling the use of new technologies.

6. Research and development are not necessary or possible in the construction industry.

7. Contractual relationships must be adversary relationships to ensure there are no conflicts of interest.

8. The construction process or system is too complex to describe or analyze accurately.

9. Unions will not cooperate in achieving better productivity.

10. There is no way to obtain productivity data and, therefore, no way to measure trade-offs to determine whether overall productivity has been changed.

Although the structure of the industry does make the resolution of these factors difficult, it is by no means impossible. There is a general agreement among many owners, engineers, architects, contractors, labor groups, economists, and others that productivity in the construction industry is a problem requiring serious study.

To provide a better understanding of how to analyze construction productivity, this chapter considers the subject under four major headings:

1. The nature of construction productivity.
2. Concepts of productivity and construction productivity.
3. Cost effectiveness in construction
4. Improving productivity.

9.1 THE NATURE OF CONSTRUCTION PRODUCTIVITY

The construction industry is frequently recognized as a bellwether of the national economy; it often precedes other segments of the private sector in cycles of increased and decreased activity. A profile of the industry in the 1970s shows why. Its expenditures amounted to approximately 11% of the gross national product (GNP); it employed 14% of the work force; and it influenced the decisions on commitments of 30 to 40% of the national resources. By the end of the 1970s, the construction industry annually spent close to $230 billion and was the largest single industry in terms of dollar volume and employment.

All of the structures and facilities required for residential, commercial, energy, transportation, industrial, agricultural, and resource management uses are built by the construction industry. The costs and time consumed for providing these facilities directly influences the downstream costs of services and products. Therefore, the effectiveness of almost all production systems and hence the national economy are dramatically influenced by construction efficiency.

The construction industry has changed considerably since the end of World War II. New materials, equipment, designs, processes, and techniques have brought changes to almost every phase of the building process. Since the early 1950s, the construction industry increasingly has depended on equipment-based technologies that assumed the major portion of the work of transporting and installing building materials and components.

Despite significant changes, construction has not experienced the same relative growth in man-hour productivity that has been achieved in other segments of the nation's industry. The lack of adequate growth of productivity has increased the relative cost of constructing needed facilities and has absorbed an expanding percentage of public and private funds that could be used in other ways.

Although improvements in productivity have long been sought, estimating actual construction productivity is difficult because of the heterogeneous conditions in the industry and the varying nature of the output or product (various sizes, functions, qualities, performance characteristics, etc. of the facilities). Measuring productivity for the construction industry for the entire country involves consideration of the many different products of different kinds of construction in different geographic regions. In addition, the technology of the product and the product itself are always changing, making it difficult to measure or improve productivity as it is done in, for instance, manufacturing industries.

The difficult measurement problems do not eliminate the real need to study construction productivity. Data are needed that will permit more rational cost comparisons among areas, to analyze the true impact of technological changes, to establish wage negotiation guidelines, and to develop strategies for reducing the costs of construction. There are also needs to (1) define areas that require research, (2) outline what management can accomplish by studying productivity and implementing strategies to improve it, (3) identify the problems and payoffs involved in all the strategies for achieving higher productivity, (4) analyze current cost indices to evaluate changes in design and construction techniques, and (5) increase industry and public awareness that construction productivity has a significant impact on the nation's economy.

9.2 CONCEPTS OF PRODUCTIVITY AND CONSTRUCTION PRODUCTIVITY

Construction is a service industry involving the transportation and processing of materials for the purpose of producing or maintaining facilities. This basic function is accomplished by taking natural and processed resources (which may be part of the site or imported to it) and changing them in form, place, and condition by the expenditure of human energy (both muscular and mental) aided by tools, power, and equipment. Many unique factors affect the movement and processing activities. Any analysis of construction productivity must begin not only with a definition of the general term "productivity," but also describe how it applies to the construction process described above.

9.2.1 Basic Principles of Productivity

Productivity by any definition or method of measurement is a comparison between input and output; an increase in productivity always means that either (1) input is reduced for the same output, or (2) the quality or quantity of output has been improved for the same input. The following is quoted from a bulletin of the Bureau of Labor Statistics (1971):

> Productivity is loosely interpreted to be the efficiency with which output is produced by the resources utilized. A measure of productivity is generally defined as a ratio relating output (goods and services) to one of the inputs (labor, capital, energy, etc.) which were associated with that output. More specifically, it is an expression of the physical or real volume of goods and services related to the physical or real quantities of inputs.
>
> A variety of plausible productivity measures can be developed, the particular form depending on the purpose to be served. . . . *No one measure is the right or best measure.*

A summary report of a workshop on construction productivity funded by the National Science Foundation states that productivity must not be defined merely as "units of measurable work from each member of the labor force." Units of work per labor hour is one gauge of productivity, but the report emphasized that any problem that ultimately affects unit production is a productivity problem.

The productivity of the construction industry can be measured in many different ways. The measure selected depends on the information required at that time. Whatever method is chosen to measure productivity, it will have to consider the four major classes of input that go into building construction: (1) management, (2) materials, (3) equipment, and (4) manpower. Output is the structure or facility that is built, or some component of the facility. Depending on the purpose of the comparison, the amount of input or output can be stated in terms of (1) quantity and quality of items, or (2) dollars, either total or per unit.

The construction industry is concerned with end products—facilities—rather than components of those products. The objectives of construction concern completed projects rather than improvements in materials or processes. From this perspective, improvements in productivity come about through (1) better management of inputs or (2) better inputs. The primary interest of the owner or contractor–builder is in the first of these, the management phase: owners and builders expect those responsible for producing the input to accomplish the second phase, improving that input. Specifically, they look to producers to improve the characteristics of materials and equipment used in construction work.

Improvements in manpower inputs are achieved in quite different ways than for other inputs. Management of manpower is a much more complicated problem, and improvements have become the responsibility of the individual worker, labor leaders, and construction management. Unfortunately (or fortunately, depending on one's viewpoint), because of basic differences in interests, disagreements arise between those responsible for management of construction and those who conduct the labor and/or management of the labor. The manpower required and the way it is used are not always optimized in terms of the available labor supply and quality.

By definition, for a given set of facility requirements, an increase in productivity means a decrease in cost. The decrease in cost is either in absolute terms (i.e., lower cost for the same products) or relative value (i.e., better quality for the same cost). Most of the discussion about productivity centers around how productivity and cost are to be measured and expressed. For the users of construction, the major consideration is unit costs in dollars. To them, a productivity increase means a reduction in, for example, dollars per square foot of structure.

A cost reduction does not necessarily mean there has been an increase in productivity. There are three basically different ways that unit costs can be reduced:

1. The imposed constraints for the project can be relaxed, for example, by lowering the standards for environmental protection.

2. The plans and specifications can be changed so that the construction is less difficult or the materials less expensive.

3. The construction process can be made more efficient through better management or improved inputs.

For the user of construction, cost reductions by any method would be economically productive and would decrease the need for investment capital. To society in general, the relaxation of constraints might be economically productive in the short run but socially and economically counterproductive in the long run.

Imposition of constraints in general (whether they involve environmental protection or worker safety or quality of material, or whatever) results in increased costs. Loosening of constraints generally results in increased productivity for a particular project. However, regulations ensuring building safety or protecting the environment are not imposed to achieve immediate economic benefits for facility users or builders. Increases or decreases in productivity related to regulation must be regarded in the light of overall, long-term economic benefits to society at large. Historically, companies have been free to choose the means to achieve greatest profit and productivity within limits; the limits depend in part on public interest. The higher costs to owners and builders have been justified in terms of overall social and economic outcomes rather than profitability *per se.*

A change in plans and specifications that reduces the construction cost is economically productive if (1) the performance characteristics of the facility are not sacrificed or (2) the sacrifice is offset by savings in construction. For example, reductions in purely cosmetic items are economically productive insofar as the facility construction is concerned. But considering the overall operation of the facility, changes may reduce the satisfaction of the workers who will occupy it. Changes that reduce the comfort or satisfaction of eventual facility occupants may be economically counterproductive even though they produce immediate cost reductions during construction.

Structural features may be modified for easier, less costly construction. Such a change would be economically productive only if structural life and maintenance or operating costs are not adversely affected. If such a change reduces the value of the facility in any way, it may be counterproductive.

Increasing the efficiency of the construction process results in a reduction in the total cost of manpower, materials, equipment, and on-site management. Provided that no change is made in either the constraints or in the plans and specifications, such savings are economically productive.

The owner/user of construction is concerned mainly with the final cost of construction. Trade-offs in cost between labor, materials, and equipment are of more interest to the construction company than to the owner.

In summary, there are a number of ways to increase productivity in construction. These involve increased efficiency in management, materials, equipment, manpower, etc.—the elements that make up total on-site productivity.

Cost changes may or may not give a clear indication of productivity changes. Reductions in cost that also reduce the quality of the facility or its environs may actually constitute a decrease in productivity. Correspondingly, an increase in unit costs that provides better performance or improved environs may, on analysis, constitute an increase in productivity.

9.2.2 ACHIEVEMENTS IN CONSTRUCTION PRODUCTIVITY

Since World War II, numerous innovations in the construction industry have increased its overall productivity. As noted before, the progress has come about through improvements in equipment, materials, and construction management. The examples listed in Table 9.1 have contributed significantly to increased productivity in the industry. Many of the items were initially developed for other industries, such as mining or agriculture. Most of the advances in construction technology are products of research and development by equipment and material producers.

Very little funding for research and development has been provided

TABLE 9.1   Partial List of Significant Developments in
Construction

Crawler tractors
Ammonium nitrate as a blasting agent
Modular building components
Earthmoving equipment
Tower cranes
Electric wheels
Articulators—loaders, graders
Concrete pumps
Down hole drills
Precasting, prestressing of structural members
Full bore tunneling machines
Scheduling and control techniques—critical path method (CPM), etc.
Construction management systems—construction management (CM) systems
Value engineering (VE)
Merit shop development
Education and training innovations

TABLE 9.2   List of Construction Productivity Information
Frequently Required by Owners/Builders

Cold temperature vs. loss in labor efficiency
Hot temperatures vs. loss in labor efficiency
Average monthly temperature conditions by month vs. various localities with
    efficiency factors for each
Productivity factors for various overseas localities vs. given standards for
    indexes (i.e., based on a 1975 standard man-hour in Reno, Nevada = 1.0)
Learning curve vs. efficiency
Altitude vs. efficiency
Lighting vs. efficiency
Noise vs. efficiency
Hours per shift vs. efficiency
Overtime vs. productivity
Packaging vs. handling vs. waste

by construction companies, labor groups, design professionals, construction users, or government. However, construction has benefited from the kind of general technological advance that all industries have enjoyed from developments such as improved lubricants, computers, metallurgy, better tires, new office equipment, and so on.

9.2.3   CONSTRUCTION PRODUCTIVITY FACTORS

Factors that affect productivity in the construction industry are usually qualitatively understood by people in the industry. Unfortunately, these factors are by no means quantitatively understood; this means they cannot be manipulated logically and scientifically at the working level. There is a need to identify the most significant productivity factors, quantify them under various conditions, and describe the relationships of these factors with each other.

In general, the industry lacks objective information on productivity factor relationships. Much work is needed not only to collect this kind of information but also to develop a system for reporting it, analyzing it, and making it available for general use. Table 9.2 gives examples of the kinds of construction productivity information often sought by owners, contractors, labor management, and others faced with problems of productivity.

The complexity and sheer total number of factors affecting construction productivity are demonstrated in Table 9.3. The major categories

TABLE 9.3 Examples of Construction Productivity Factors by Major Category

Management

Planning (particularly preplanning)
Scheduling
Organization
Supervision including dilution
Communication
Human Factor—worker motivation—working conditions
Site layout
Material locations
Heights
   Work area from floor
   Work area from ground
Information systems
Size of organization
Inspection/quality control
Finance/budgeting
Safety
Project factors
   Duration
   Size
   Type construction
   Type contract
Purchasing and delivery practices
Productivity bargaining

Technological

Materials
Methods
Equipment

Regulatory

Government (all levels)
Codes and standards
Social legislation
Bureaucratic structure

Laborer

Motivation
Overtime/fatigue
Work rules, including restrictions on labor-saving tools
Training and skill improvement
Availability/supply/age

TABLE 9.3   (*Continued*)

Attitude
Crew size and composition
Learning curve
Trade stacking
Worker know-how
Acceptance of change
Flexibility
Absenteeism
Tardiness
Turnover
Mobility
Jurisdictional disputes
Temporary work assignments
Length and number of shifts
On-site travel

Engineering/Design

Standardization
Uniformity
Constructibility
Innovation
Errors/omissions
Type contracts
Lack of R & D allowance

Craftsmen

Construction techniques
   Manipulative skill
   Division of labor
   Mechanical aids
   Prefabrication
Physical properties of materials
   Unit size
   Quality of workmanship and materials
   Modular coordination
Weather
   Prevention of work
   Weather protection
   Damage to work
Training
   Teaching methods to reduce motion and increase skill
Labor–management relations
   Work rules

TABLE 9.3   (*Continued*)

Worker motivation
  Worker satisfaction
  Incentive systems
Motion and time study (for improvement)
  Efficiency
  Less difficult
  Less fatigue
  Work smarter
Site control
  Craft coordination—composite crews
  Scheduling
  Material supply
  Equipment
  Horizontal and vertical transportation of material
  Safety requirements

Other

Delays
Changes
Weather/climate/seasonality
Economic conditions
Altitude
Noise
Lighting
Safety (OSHA)
Owner budget
Nonuniformity in building codes
Quality vs. production

of factors are (1) managerial, (2) technological, (3) regulatory, (4) labor related, (5) engineering/design, and (6) craftsman related.

As is discussed in a later section, the decision-making hierarchy in construction further complicates the study of productivity factors and their interrelationships. For decision makers at all levels it is the case that some factors may be controlled but some are beyond control.

In regard to manpower, the problem of determining and measuring productivity factors is even more complex. These factors have been the subject of numerous general studies and publications on employee productivity. An authorative source of explanation and information is Sutermeister's *People and Productivity* (1976).

9.2.4   STUDIES ON CONSTRUCTION PRODUCTIVITY

The current body of knowledge concerning productivity and related research areas is listed in several recent bibliographies (Taylor, 1978; Construction Industry Research, 1980). There are many internal reports and a lot of excellent data in the files of individual firms and agencies. Productivity figures (bricks laid per day, lengths of pipe laid per hour, etc.) are often treated by construction firms as are other cost figures—as confidential. Some research is available on areas such as labor management.

The types of industry productivity data that are available are illustrated in Figure 9.1 and Table 9.4. There is general agreement that

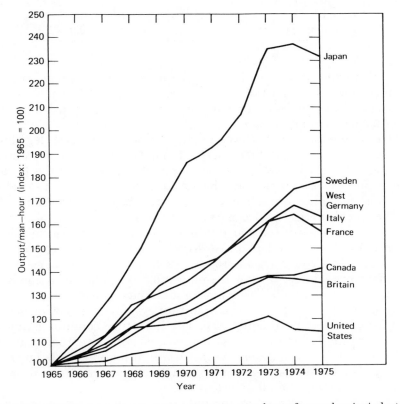

Figure 9.1   Comparison of productivity output per man-hour of several major industrial powers over the 10-year period 1965–1975.

TABLE 9.4   Annual Percentage Rate of
Productivity Change for Selected Industries, in
Terms of Output/Man-Hour, 1958–1969

Industry	Productivity change (%)
Cathode ray picture tubes	11.6
Electronic computing equipment	8.6
Railroad transportation	6.3
Motor vehicles	4.1
Cigarettes	2.1
Residential construction	2.0
Nonresidential construction, except highways and sewers	1.5
Highways	1.0
Bolts, nuts, rivets, washers	0.3

overall productivity growth in the United States has not kept pace with the growth in most of the other major industrial nations. Within the United States, construction productivity has lagged behind that of other major industries.

Overall productivity has been greatly influenced by social legislation related to environment, labor, occupational safety, consumer interests, etc. Such legislation can increase construction costs and the time required for project completion. The effects of regulations and standards established by law are schematically shown in Figures 9.2 and 9.3.

The cost of energy has become an increasingly critical social problem. Costs of power-plant construction have also become more significant. Figure 9.4 shows power-plant construction costs increasing over time. For large-scale construction projects, such as power plants, good management procedures, worker training programs, and reduction of absenteeism, for example, are major productivity considerations.

There is a scarcity of information that systematically addresses all the factors affecting productivity. The literature contains many references to single factors (e.g., motivation, overtime, weather, materials delivery, adequacy of plans). Very little information exists on the effects of combinations of factors. The following paragraphs illustrate how these factors can affect productivity.

The adverse effects on productivity of excessive use of overtime are illustrated in Figures 9.5 and 9.6. These data were obtained for elec-

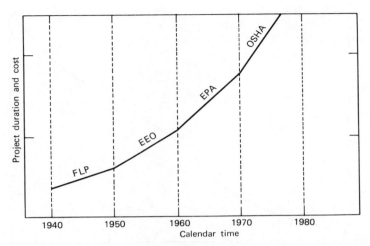

Figure 9.2 The nature of the cumulative effect of regulations and standards resulting from social legislation eras in the United Stated from 1940 to 1980.

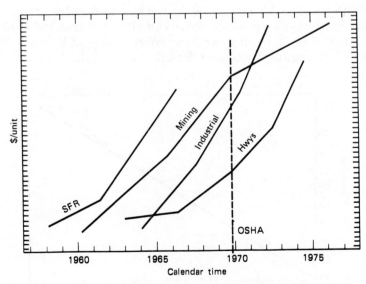

Figure 9.3 The nature of the effect of regulations and standards on unit costs in different areas of construction in the 1900s.

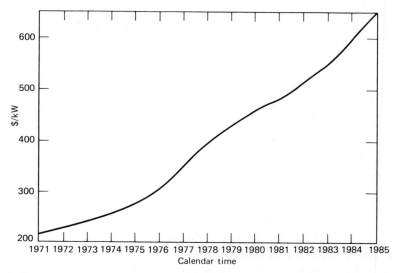

Figure 9.4 The cost trend for thermal power plants from 1971 to 1985 in dollars per kilowatt.

trical workers. The trends show significant losses in worker productivity with increased overtime hours worked week after week.

Climatic effects also produce serious on-site problems, reducing both the number of working days and the effectiveness of workers. Tables 9.5 and 9.6 illustrate the type of data that are available on the effects

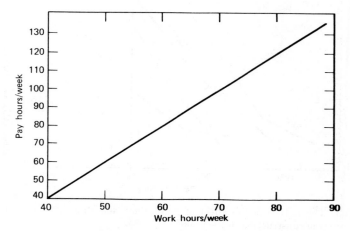

Figure 9.5 The effect of overtime on cost of labor.

TABLE 9.5 Reduction in Work
Efficiency at Hot Temperatures
from Standard Efficiency at 70°F

Effective hot temperature (°F)	Loss in efficiency (%)[a]
80	30
90	40
100	60

[a] Data are for trades using gross skills: laborers, concrete handlers, cement masons, ironworkers, operating engineers, roofers, bricklayers, glaziers. No estimates available for trades using fine skills.

TABLE 9.6 Reduction in Work Efficiency at Cold Temperatures from Standard Efficiency at 70°F

Effective cold temperature (°F)	Loss in efficiency (%)[a]	
	Trades using gross skills[b]	Trades using fine skills[c]
40	0	15
30	0	20
20	0	35
10	5	50
0	10	60
−10	20	80
−20	25	90–90+ (probably can't work)
−30	35	

[a] These data are based on the assumption that the worker has the proper clothing to provide cold protection.
[b] Examples of trades using gross skills are given in Table 9.5.
[c] Examples of trades using fine skills are carpenters, tilers, plumbers and pipe fitters, welders, and electricians.

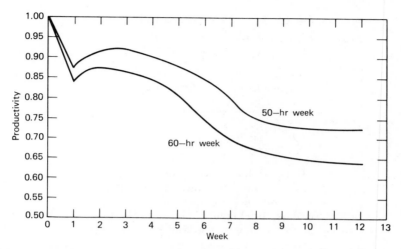

Figure 9.6 The cumulative effect of overtime on productivity for 50- and 60-hr work weeks.

of temperature on work efficiency. Construction workers are also influenced by environmental factors such as altitude and light. Figures 9.7 and 9.8 illustrate the adverse effects of these factors on productivity.

The importance of good construction management is illustrated in Figures 9.9 and 9.10. Careful planning and organization can increase

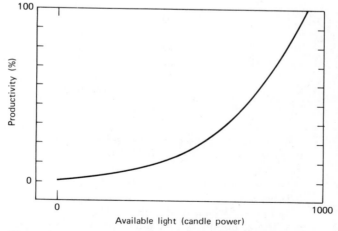

Figure 9.7 The effect of light variation on construction productivity.

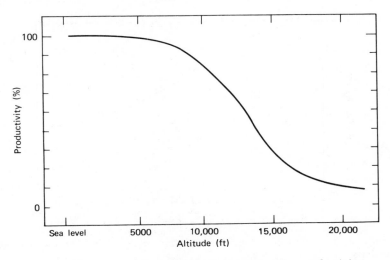

Figure 9.8   The effect of altitude on construction productivity.

manpower productivity; simply decreasing the distance between the work area and required materials can increase manpower effectiveness. Another management consideration is how time is spent by workers on the job. Figure 9.11 shows a breakdown of worker time on a typical large power-plant construction project. Such information can

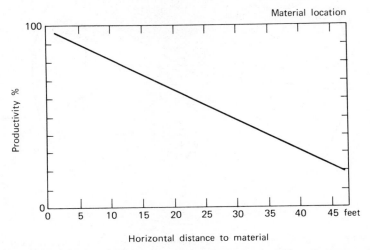

Figure 9.9   The influence of horizontal distance from material location on construction productivity.

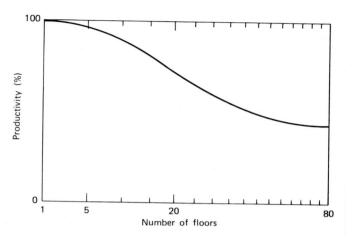

Figure 9.10 The effect of vertical distance from material location on construction productivity.

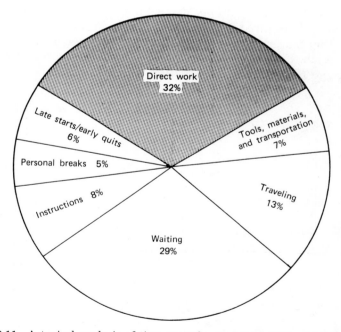

Figure 9.11 A typical analysis of time spent by workers building a nuclear power plant.

298

lead to better organization of work so that more time is spent on direct work by each member of the crew.

9.3 COST EFFECTIVENESS IN CONSTRUCTION

A cost-effectiveness analysis is a study of alternative methods to achieve an economic objective. The purpose of analysis is to determine either (1) the lowest cost of a facility for a given set of facility requirements or (2) the best facility that can be obtained for a given investment.

9.3.1 PRINCIPLES OF COST EFFECTIVENESS

Construction costs are a function of three interrelated sets of factors: (1) facility requirements, (2) plans, specifications, and constraints, and (3) on-site factors such as manpower, materials, services, equipment, and management. The facility requirements establish an irreducible minimum cost. These requirements are manifested in plans and specifications for the facility as well as in some of the constraining conditions under which the construction must be done. For any one project, the constraints, plans, and specifications dictate an irreducible minimum cost involved in the use of some combination of manpower, materials, services, equipment, and management that will accomplish the job. Construction costs for a given facility can be reduced only by changing one of these factors.

Constraints consist of institutional and physical conditions. Institutional limitations are imposed by federal, state, or local governments, and by the practices of engineers, administrators, labor management, and business management. They can be changed only by techniques appropriate to the type of institution in question.

Physical constraints include those imposed by geology, climate, geography, and other factors that cannot be changed. Effects of physical constraints can often be modified; this usually results in increased costs (e.g., work in hot environments may require air conditioning of equipment cabs and work in earthquake areas may require extensive adaptation of the construction process).

On-site costs of manpower, materials, equipment, and management can be changed in an almost infinite number of ways and still meet the requirements of constraints, plans, and specifications. If dollar costs are to be the only criterion, the minimum cost is sought by various adjustments. Such adjustments may well lead to higher costs for one item that produce greater savings in another area. For example, if equipment costs were increased by $100,000, enabling manpower costs to be reduced by $500,000, a net reduction in total cost of $400,000 would be achieved.

Assuming that the objective is to produce the largest reduction in dollar cost of construction, analyses can be made of the magnitude of the savings versus the investment cost to achieve that savings. If costs or benefits are in nonmonetary terms (e.g., a reduction in pollution), a subjective decision is required relative to the benefits derived. If the cost reduction is in terms of unit costs (e.g., dollars per square foot of floor space), the dollar reduction will have to be considered in terms of total dollars or total square feet of construction for the project.

The evaluation of alternative ways to reduce costs consists of more than the benefit–cost analysis. A primary consideration is feasibility; the following questions, at least, must be answered:

1. Is it economically feasible—are the investment funds available?
2. Is it socially feasible—will the method be socially acceptable?
3. Is it environmentally feasible—will the consequences be acceptable in terms of environmental effects?
4. Is it politically feasible—can political support (in or out of government) be obtained?
5. Is it technologically feasible?

Another factor to be considered is time, that is, how long it will take to achieve the cost reduction. The solution may require too much time for the probable benefit.

From a broad consideration of construction cost reduction, it has become clear that a process of *suboptimization* must be avoided. That is, a major change or "improvement" in one element of the process may seem simple and attractive. But it may be less effective in reducing overall costs than modest improvements in several inputs. In fact, a change may have no effect on reducing overall costs. The entire system of construction inputs and outputs must be examined to decide what kinds of changes will be cost effective. Too often, big changes are made that have deleterious economic effects that could have been predicted if the relations between the various elements of the project had been examined.

The users of construction can be more objective about cost effectiveness than can other people in the industry. Users are primarily interested in cost reduction, and much less interested in how the reductions are achieved. A cost-effective solution is one that produces the desired facility at a lower cost than alternative methods.

9.3.2   ORGANIZATION OF A STUDY

If the objective is to reduce the costs of construction to the lowest possible figure, the sources of costs must first be identified. For this purpose, data such as those in Figure 9.12 provide perspective on total expenditures and major items of cost for the industry or sectors of the industry.

Two basic controls on costs are the project requirements and the constraints. The former constitute the production and operational needs of the facility to be constructed. Constraints include investment limitations, building codes, and regulations. Changing either of these results in changes in the cost of the facility.

Two broad classes of construction activity that determine costs both

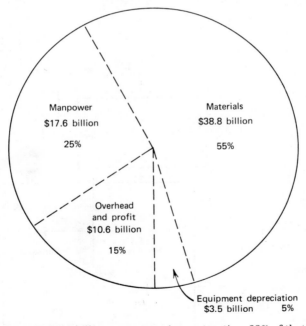

Figure 9.12   In 1980 $223 billion were spent in construction. 32% of that amount was used in industrial, commercial, and utility construction. The figure analyzes how costs for industrial, commercial, and utility construction can be divided into subcomponents. (SOURCE: U.S. Dept. of Commerce, 1979.)

directly and indirectly are (1) preconstruction activities and (2) actual construction work. Preconstruction activities include all of the administrative work from project conception to the start of construction; a certain direct cost is involved in this. Other preconstruction activities markedly influence ultimate construction costs by creating delays that produce (1) production losses for the ultimate users of the facility and (2) construction cost increases owing to inflation.

The major controlling factor in construction costs is the set of plans and specifications. Costs are influenced by the structural requirements and by the constraints on construction work imposed by the specifications. Direct construction costs are those already identified: manpower, materials, equipment, and management costs.

Significant reductions of direct construction costs could result from changes in regulation, technology, management, or manpower. Changes in governmental codes and regulations impact on cost. New technology may improve productivity and reduce overall cost, or it may not. Improved project management at every stage in the process can cut costs; poor management can certainly increase them. Labor supply and man-

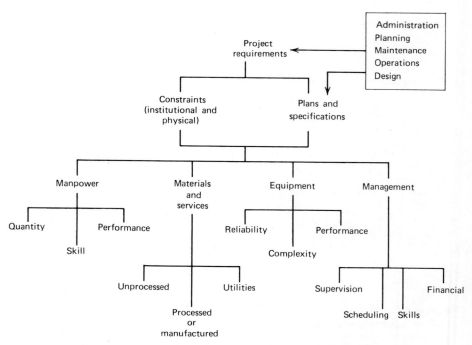

Figure 9.13   Relationships between sources of construction costs.

a. Analysis matrix for potential reductions in construction costs

Opportunities or problems	Manpower	Materials	Equipment	Project management	Total
1.					
2.					
3.					
4.					
Total					

b. Analysis of potential targets for reductions in construction costs

Activity	Manpower	Materials	Equipment	Project management	Total
Labor effectiveness					
Labor supply and training					
Project management					
Construction technology					
Codes and regulations					
Total					

Figure 9.14 Simplified format for analysis of total costs of construction innovation.

power effectiveness feed directly into costs. Figure 9.13 illustrates the connections between various sources of cost.

Although much of the preceding has application to industry-wide productivity, the principles can be applied to local problems by specific companies or groups. Forms illustrated in Figure 9.14 are typical of aids for cost-effectiveness evaluations. Such evaluations should be done for any proposed change in operation or materials. Note that *total* optimal reduction in cost is the goal in every case rather than suboptimal cost reductions in single elements. Cost-effectiveness evaluation is a fine tool for selecting the lowest-cost alternative.

9.4 IMPROVING PRODUCTIVITY

As has already been pointed out, recent improvements in construction productivity have resulted primarily from advances in equipment and materials and, to some extent, from occasional use of new analytical techniques by management. To obtain significant increases in productivity from this point on, data must be collected that can be used to identify optimally cost-effective construction methods. Some of the newer approaches and concepts relevant to increasing productivity are described in the following subsections.

9.4.1 ANALYSIS AND MEASUREMENT OF CONSTRUCTION PRODUCTIVITY

There is, as yet, no single satisfactory way to measure productivity in construction. The industry is too varied and specialized; productivity depends on a vast number of interrelations between trades, contractors, designers, owners, and so on. No universally accepted indexes of construction productivity exist. A good index of productivity requires (1) an accurate gauge of the value of construction output and (2) an accurate assessment of manpower output.

Productivity comparisons are made at different times; today's level is compared with that at times past to evaluate whether productivity is increasing or not. This means that measurement is complicated by fluctuating currency values and technological changes that occur over time. A useful measure of productivity requires an accurate assessment of change in output. In construction, the output is not a single, easily counted product but a complex product with numerous and different components. The most convenient way to measure construction output is in terms of value.

For manpower, no effective system exists for determining the relative productivity of the trades or the alternative construction processes that are used. Manpower productivity per hour or per day can be measured by time-and-motion studies. A significant advantage of time-and-motion studies over other methods is that they generate uniform data that enables comparisons within a craft of various tasks and processes. A major limitation is that time-and-motion studies cannot be used to compare the productivity of different operations. A time-and-motion study of a laborer digging a ditch may show he is working 90% of each hour and producing 1 yd^3 of excavation per hour. A similar study of a backhoe operator may show him to be working 90% of each hour but producing 100 yd^3 of excavation per hour. The studies cannot determine which man is more *productive*. Both are working 90% of each hour but the cost of excavation per cubic yard of the two operations is very different (since the cost for the heavy equipment is very high).

The construction industry needs a simple, unified system for assessing productivity. The system should be capable of measuring the productivity of any construction operation so it can be compared with that of other, alternative operations. A construction planner or manager should be able to use the system to identify the least productive aspects of the project under construction. If these least productive links can be identified consistently, each construction company and the industry could achieve substantial improvements in productivity.

9.4.2 USES OF PRODUCTIVITY MEASURES

The reason for measuring productivity is almost always to improve it. The process of improving productivity must take place at many levels of the construction process. The productivity question must be asked every time a decision is made, from start to finish on a project. The management must consider the issue of productivity as a matter of course in planning and implementation at every stage and level of work. Productivity measures must not be left for last-minute use by a superintendent on the job after something has gone wrong. It cannot be sufficiently emphasized that productivity must be systematically considered from the outset, the initial conceptualization of a project. If this is not done, not much can be done to improve it at the bottom line.

A big problem in the industry is that there is no general information feedback system. Information about on-the-job productivity gains from one project or context does not become available for others to adapt for their own projects. The development of a central body of production information may well be a more critical problem than any other. Although the actual, on-the-job productivity measurements provide answers to the question of improved production practices, they will be useless to the industry unless the information is widely disseminated and available to all planners, managers, and decision makers.

On-the-job labor costs are often the principal area where significant savings can be realized by increased productivity. Work sampling and time-lapse photography are proven methods for assessing on-the-job productivity. These and several other methods are discussed in Section 9.4.3.

Studies of construction productivity can rarely use the same methods as studies on, say, manufacturing production. Unit work rate on a construction job varies because of continuous change as the work progresses. Output is influenced by the situation at the job site at the moment, which is different from what was true last week or at some other time. This points up once again how scheduling and planning of interdependent events have a major effect on productivity.

Field data collected to assess productivity must be complete and detailed to provide the basis for valid comparisons. Obtaining good field data requires cooperation of all concerned parties; it is a trust that must not be abused. Workers will cooperate in the interest of overall improvement but not when the results of a study are used to embarrass individuals or as a basis for punitive actions.

The type of unit used in assessing output depends on the requirements of the analyst. A unit that is appropriate to the total construction effort must be chosen. Each type of measurement unit has some advantages; no one measure is best for every analysis or every purpose. Contractors, for example, are interested in units that can be used for estimating, budgeting, job control, etc. Consequently, they require convenient units such as (1) output per man-day, (2) output per man-hour, (3) output per dollar value of the total labor, (4) man-hours per dollar of construction cost, or (5) man-hours per dollar of contract labor cost. These units may or may not be usable or meaningful to other interested parties in the construction process.

9.4.3 METHODS OF MEASUREMENT OF ON-THE-JOB PRODUCTIVITY

In the following subsections, five typical methods of measuring productivity are discussed. They are (1) work sampling, (2) time-lapse photography, (3) cost estimates, (4) key activity measurement, and (5) ratings.

Work Sampling

Work sampling is a method of assessing contractor or worker job performance rather than work accomplished. It provides a "snapshot" of the job, not a continuous picture. The method identifies weaknesses or problems and is best suited for analyzing productivity on long-term jobs. Work sampling depends on gathering enough data for statistical analysis; short-term projects may not provide sufficient data for numerical treatment and may be completed before results become available.

The method consists of categorizing the activity observed in a particular work area at a particular time. An observer classifies the workers' behavior into relevant categories such as working, personal, transport of materials, waiting, and communication related to the job. More elaborate categorizations may be useful. Work locations must be mapped by the contractor according to the phase of the project. To obtain representative samples of each kind of work, sampling must be carefully scheduled at every time of day, day of week, phase of project, and so on. It is desirable to rotate observers so that each crew of workmen is studied by more than one observer. The study should be set up so that four to six observers work in teams of two.

Work sampling is a relatively simple way to identify sources of delay, "goofing off," and inordinate transport time. However, it cannot provide the basis for comparison of work accomplished and it cannot be used to distinguish effective work from busy work. Other advantages and disadvantages are listed below.

Advantages

1. The method enables the analyst to develop a good feel for the whole job.
2. If continued from start to finish of the on-site work, the results will show when and where the productivity problems arise. For example, work sampling may show that the beginning and end of a job are more productive because, in the middle period, factors such as congestion, slow deliveries, or inability to use machine pacing have slowed things down.
3. Work sampling can be used immediately to spot problems that need instant correction. An example is the case where the elevator man was not paid overtime. Since he wanted to quit at the end of his shift, the crews had to start down much earlier than necessary.
4. The method establishes good habits on the construction project.

Disadvantages

Used in its simplest form, the method makes no distinction between effective and ineffective work. Nonproductive or low-quality work and rework are not separated from effective work. A more elaborate method by which these distinctions can be made is presented by Parker and Oglesby (1972).

Time-Lapse Photography

This method consists of sampling by camera over time. By focusing on critical operations of high cost and low productivity, this measure can help resolve problems by demonstrating their exact nature. However, it is a time-consuming and expensive process. Therefore, the target areas for time-lapse study must be carefully selected for maximum payoff.

Time-lapse photography studies must employ systematic sampling of the work motions and events of interest. When good data are collected, a detailed productivity analysis can be performed on the operation photographed, and corrective information may be provided.

Advantages

1. A detailed picture is provided of actual construction operations. This picture can be used to identify "busy" work and low-quality efforts. It can also be used to identify the most effective work practices.

2. A record is made that can be analyzed any number of times for any purpose.

3. The equipment can be set up and left to operate. It does not require a human observer.

4. The recorded data can be reviewed by the work crew and management. This can lead to immediate task improvement and suggestions for beneficial changes.

5. The opportunity for workers and management to view the films together has proved a very satisfactory way for the people involved to meet and discuss problems.

Disadvantages

The method cannot cover the whole construction project. Camera placement determines what information is recorded; important information is lost if an inadequate placement is made.

Cost Estimates

The estimated cost of construction can be used in combination with the cost reporting systems of a company to provide a built-in method for assessing productivity as the project progresses. The first step in such a system would be to make the estimates completely compatible with the actual construction operations on the project over time. Obviously, if the estimates are not based on construction activities as they are actually performed, there can be no valid comparison. Likewise, the cost reporting system should be a crew-based system of the same type as will be in effect on the job. If both the estimate and the reporting system directly reflect the on-the-job, day-by-day realities of the work, then they can provide a good measure of project productivity.

In many construction companies, this is never the case. One of the main reasons is that the pressure of competitive bidding and the tremendous difficulty of conceptually "building" a job require that the estimates be approximate. These educated guesses are rarely good yardsticks for actual productivity. The problems in using this system to measure productivity are most apparent when the estimate was much lower than actual construction costs. It is impossible to tell whether the estimate was unrealistically low or productivity was much lower than anticipated.

Advantages

The built-in processes of construction cost estimating, cost reporting, and cost control are used.

Disadvantages

1. It is very difficult to make good cost estimates. There are too many imponderables in actual construction.
2. When estimated cost is lower than actual cost, it is impossible to tell by this method whether the estimate was too low or productivity inadequate.

Key Activities

The manpower costs for certain key activities which occur in all construction jobs are assessed. Examples of key activities are installing pipe or wire. The key activities chosen are considered to be representative of all activities in the project. The concept is that the productivity levels of tasks which are common to all construction jobs can be considered representative of productivity on a particular project.

The validity of this assumption undoubtedly varies. Each company should test the assumption by studying available data on completed projects.

Advantages

1. If the key-activity assumption is met, then the procedure can compare the productivity of different projects that are widely separated in space and time.
2. This is a quantitative measurement and is relatively easy to use.

Disadvantages

1. The method provides a yardstick which may *look* more reliable than it is. Laying pipe on one project, for example, may be a particularly difficult process for reasons of design, safety, or physical or climatic constraints. So this "key activity" should not be compared with productivity on a site where laying pipe was fast and easy.
2. The assumption of direct correlation between the easy-to-measure costs of a project and all the other costs may not be valid in many cases.
3. To concentrate attention on one segment or activity may distort management's concern for the total job.

Ratings

Management productivity may be assessed on the basis of a rating given to the construction supervisor. A study performed at Stanford University (Samelson, 1977) measured the following characteristics of construction foremen: ability to meet costs, productivity level, ability to work under pressure (rescue a bad job), and administrative ability. Each of these characteristics were rated on a four-point scale: "excellent," "above average," "average," or "needs help in this area." The method has also been used successfully to measure the productivity of project managers and field superintendents (Hinze and Parker, 1978). It is a useful tool for comparing managers.

Advantages

1. The measure is easy to obtain.
2. Several aspects of productivity are taken into account.

3. Emphasis is placed on management rather than manpower productivity. It is clear that crew and project productivity depend on good management.

Disadvantages

The judgments of different raters vary considerably. This makes it particularly difficult to make comparisons between companies (since each company will have a different individual rating the field managers).

9.5 THE HIERARCHY MODEL OF THE CONSTRUCTION SYSTEM

Applying cost-effectiveness logic to construction productivity problems requires that the many factors in the overall construction system be recognized. No change can be beneficial if it is made without due consideration of all the factors that will be affected. Changing one part in any large system means changing the way other parts will function. There are a number of procedures and models to deal with this problem. The hierarchy model for the construction industry (Kellogg et al., 1978) is discussed here.

The hierarchy model describes the principal elements of a construction system and places them in perspective. Major components of the system are examined in terms of how their effects can be measured to enable cost-effective decisions.

9.5.1 ASSUMPTIONS OF THE HIERARCHY MODEL

People in the construction industry have assumed productivity level to be determined solely by construction operations on the job site (square feet of forms per carpenter hour, pounds of reinforcing steel installed per ironworker hour, etc.). These easily measured on-site factors are important elements in the overall productivity equation but they are not the only crucial elements in it. The interrelations of *all* the inputs to a project, from its very inception, also figure into overall productivity. Overall productivity is most usefully and comprehensively stated in terms of utile value (e.g., dollars per kilowatt of generating capacity, dollars per square foot of family housing, dollars per mile of interstate highway). To analyze and effectively attack the problem of productivity, the industry needs a model that includes all the elements of production integrated into an expression of the final dollar cost of the delivered product. The model must describe the construction hierarchy in terms of its component parts and their interactions.

Measuring productivity for the national economy is difficult; it is no less troublesome for construction. As previously mentioned, the industry is vertically and horizontally fragmented. No single firm controls as much as 1% of the industry's gross product. This fragmentation makes it difficult to bring the industry's resources to bear on any problem, however crucial, including the development of tools and procedures for increasing productivity. Assembling meaningful information from the thousands of different sources is virtually impossible.

A model of the entire process of construction, from project conception to delivery, would be useful for analyzing construction productivity. The model must address the issue of productivity at every level of the industry and phase of work. Most important, the model must be useful to all elements of the industry.

The hierarchy model addresses the productivity issue at five industry levels: (1) policy formation, (2) program management, (3) planning and design, (4) project management and administration, and (5) on-site construction. The model is not proposed as a finished product or as a "canned" program to be plugged into a computer. Rather, it is a conceptual framework—a method of understanding a complex system. The hierarchy model can be used by various sectors in the industry. It also serves as a starting point for rational decision making and judgments at any level. The salient features of the model's assumptions are listed below.

Effectiveness, not efficiency, is the major criterion. Increased productivity is obtained by working smarter, not necessarily harder. People generally want to produce and to feel productive. Productivity is maximized by establishing a climate for effective operation, taking into account all elements of the work process.

Maximizing overall productivity of a project is the goal. It is more important than achieving high productivity in any of the component parts. High unit productivity can destroy working relationships and may contribute little to overall productivity. Management decisions not normally considered as affecting productivity may severely limit overall productivity.

Productivity can be enhanced by controls. However, owing to the extreme variability of the construction environment, it is difficult to transfer, duplicate, or even maintain control systems over time. Interpersonal working relationships may be more important in achieving transferable productivity gains.

Productivity improvement may not be transferable.   Results for one type of construction or even a whole job may not apply to another. Qualitative comparisons of factors contributing to higher productivity are probably more generalizable than quantitative measures.

Productivity improvement has to be a part of the total management system.   Viable mechanisms for information on productivity to be delivered and used must be established.

Some Productivity measurement methods may be adapted from the manufacturing industry.   These methods can be tailored to the job under study to obtain meaningful data.

9.5.2 DEFINITIONS

The hierarchy model attempts to break the constraints imposed by traditional industry definitions. A nomenclature developed by economists proved suitable for describing the levels of the construction industry:

"Macro" levels: involving large quantities of considerations. The macro levels are (1) policy formation, (2) program management, and (3) planning and design.
"Micro" levels: relating to a small area or consideration. The micro levels are (1) project management and (2) on-site construction.

The terminology is not meant to imply that one level is more important than others in determining productivity. The terms merely distinguish the scope of consideration of each level.

Table 9.7 illustrates the hierarchy model as applied to the example of highway construction. At the "policy formation" level are the policy considerations of the federal government, other national forums, and major state and regional governments. The primary impacts on construction productivity at this level appear in terms of market (in what sectors will funds be invested?) and regulation (environmental constraints, resource allocation, and so on). The impact of decisions and actions at the policy formation level can be enormous. Consider the following:

1. For a new coal-fired power plant, the cost of complying with federal and state regulations amounts to 45% or more of the total cost.

2. To develop a new mining property on federal lands (or for any other project using federal funds) it can take three or more years to obtain final approval from the Council on Environmental Quality. Construction costs escalate at about 15%/year.

TABLE 9.7 Hierarchy Model of Construction Productivity for Highways and Streets[a]

Policy formation	Billion $	Program management	Planning and design	Project management administration	On-site construction
Private Construction					
Residential building	80.4				
Nonresidential building					
Industrial	7.0				
Commercial	13.9	*4000 Miles of Interstate*			
Hospital and institutional	3.3	System			
Other	2.8	$2,000,000/mile			
Total nonresidential bldg.	27.0	Geographic location			
Form construction	2.5	Number of lanes			
Public utilities		Number of bridges	*Concrete Bridges*		
Telephone and telegraph	4.0	Terrain	$250,000/mile		
Electric light and power	11.0	Location environment	or		
Other	5.6	Alignment	$25.00/ft² of bridge		
Total public utilities	20.6	Line of site	Financial	*Design*	
All other private construction	1.0	Urban	Design	$200/ft² of bridge	
Total private construction	131.5	Suburban	Land	Planning	*Survey*
Public Construction		Safety	Contract administration	Survey	$0.20/ft² of bridge
Buildings			Site work	Specifications	$/MH/control points/EO.
Educational	5.2		Substructure	Structure analysis	$/MH/layout/unit
Other	7.0		Excavation	Budgets	$/MH/as-built location
Total buildings	12.2		Steel bearing pile	Drafting	
Highways and streets	9.6		Resteel	Footing concrete	
Military facilities	1.6		Concrete	$3.00/ft² of bridge	
Conservation and development	3.7		Backfill	*Labor*	
Other public construction			Superstructure	Expendable materials	*Labor*
Sewer systems	5.8		Columns and caps	Permanent materials	$1.00/ft² of bridge
Water supply facilities	2.0		Deck	Equipment	$/MH/forms/ft²
Miscellaneous	3.2		Approaches	Overhead	$/MH/concrete/yd³
Total other public construction	11.0		Guardrail	G/A	$/MH/finish/ft²
Total public construction	38.1			Profit	
Total new construction	169.6				

[a] MH = man-hour. G/A = General and Administrative.

3. In 1965, construction of a nuclear plant required about 3 million pages of documentation. By the end of the 1970s, more than 20 million had to be amassed.

Referring again to Figure 9.18, the "program management" level is the level of policy interpretation. Specific programs and criteria are developed and the scope of projects is defined. Here, program managers consider the productivity impacts of design and quality criteria established by the government and by professional standards associations. It is at the program management level that intangibles such as esthetic values, the public good, quality of life, and acceptance of risk tend to blur the impact of productivity figures.

People at the "planning and design" level interpret the project in terms of specifics. To a significant extent, they determine the cost of physical construction through choice of materials, date of commencement, acceptable construction processes, personnel, and the facility requirements. In terms of total productivity, the success or failure of a project is often determined at this level.

At the "project management and administration" level the range of choices for improving productivity is decidedly narrowed. The project management makes choices of suppliers and contractors, does the detailed scheduling and planning of labor and equipment use, and makes other decisions which have only superficial effects on total productivity. The sooner in the life of a project that people at this level become involved, the more opportunity there is for improving productivity.

Finally, at the "on-site construction" level, only effective and intelligent employment of labor and equipment offers any hope of improving productivity as the work progresses. Although work rules, working conditions, skill utilization, and the like may be improved, the state of the art in these areas is quite advanced. Further increases in productivity at this level are likely to be marginal.

9.5.3 USING THE HIERARCHY MODEL: AN EXAMPLE

Table 9.7 illustrates the application of the model to the decision-making hierarchy of the interstate highway construction system. The figure illustrates the concepts as they could be applied to any other sector of the industry.

At the policy formation level (left-most column), the construction industry is divided into sectors. Total dollar volume in each sector is initially determined at this level. Through a variety of means, government policy makers can influence what is spent in each sector. For example, more stringent environmental regulations increase spending on utilities construction; lower federal interest rates increase output in the housing sector; lower depreciation rates and lower investment credit retard industrial and commercial building. In the case of interstate highway construction, the decision by Congress to fund the interstate highway system allocated federal money to this use rather than to improving the network of local and state highways in existence. Because of this policy decision, less federal money was available for competing projects.

The program management level decisions are illustrated in the second column of Table 9.7. For example, the decision to use bridges instead of following natural terrain probably decreased the overall construction productivity to obtain the goals of increased speed limits (which are now illegal) and improved safety. As mentioned earlier, the considerations at this level are often intangibles such as these. The trade-offs cloud the productivity issue.

The example of the bridge decision is continued in the third column of the figure. Planning and design level decisions are those made by state highway departments and architects and engineers. They determine the materials to be used, configurations of structures, selection of components, and overall schedules for beginning and completing the various projects. Many times these decisions are made with little regard for implementation or execution problems. In the bridge example, the regional availability of concrete versus structural steel could be an important factor in determining final cost and, frequently, such questions are not considered at the planning and design level. This kind of oversight makes it impossible for other levels to meet cost and time estimates. Other questions may also be critical. What construction crafts have more work scheduled than is possible during the planned period of the projects? Can the design or schedule be altered to make the need more uniform for crafts in short supply? These kinds of analyses must be made at the planning and design level as well as decisions concerning the usual planning tradeoffs.

At the project management and administration level, engineers and contractors analyze the logistics for the project and secure the necessary labor, materials, permanent equipment, and construction equipment. Detailed work planning takes place in conformance with the specifications, plans, and schedule requirements.

The objective at the project management and administration level is to provide the right labor at the right time, equipped the right way to install the right materials in the most effective sequence. This is an admirable goal but one that is seldom achieved. However, decision makers at this level do the best they can given the constraints already "cast in concrete," to bring the job in on time and within the budget. Effective project planning and administration within these constraints is their important contribution to total productivity.

Finally, at the on-site construction level, construction superintendents, foremen, union stewards, and craftsmen make multitudes of individual decisions affecting unit productivity. Perhaps here, more than at any other level, effective interpersonal relationships play a major role in productivity. A cooperative, enthusiastic work environment can ensure that unit productivity contributes to total project productivity. An adverse, strife-ridden work environment can create a complete breakdown in production. Because of the narrow range of decision alternatives at this level, a negative contribution is far more likely than a positive one. In other words, a well-executed site effort will meet the owner-established cost and schedule goals, but rarely exceed them. A poorly executed site effort, or labor inefficiency, can only increase the costs and lengthen the schedule.

The importance of the hierarchy model is in its ability to describe all the elements and factors that go into the ultimate productivity expression (in the example, productivity would be expressed in dollars per highway mile). Models may be devised for all of the major construction end products. Then the common elements can be defined and like operations can be compared from one product to another. That is, the cost per square foot of concrete forms can be compared whether they are for bridges, hydro power houses, commercial buildings, institutional buildings, or thermal power plants. The key to the analysis is the reduction of the construction product into its basic components so that the productivity factors can be identified and analyzed. This minimizes the effect of industry fragmentation on the assessment of productivity.

The industry's basic decision-making processes have not been sufficiently visible or understood. Effects on productivity have not been adequately recognized. In many cases, ultimate project productivity is determined by decisions on alternatives, by who influences the decisions, and by the decisions of what, when, and where to build. It is therefore essential to describe the complete construction process—how the product emerges from the early decision to build down to the final placement of each square foot of wall board.

To facilitate the process of clarifying the decision-making process, the "productivity background data matrix" was devised. Table 9.8 shows the use of the matrix on the complex example of the interstate highway system. The matrix identifies the basic considerations and

TABLE 9.8 Productivity Background Data Matrix for Highways and Streets[a]

Basic considerations and alternatives		Basis of decision	Influence on decision	Implementor	Productivity factors	Measurement basis	Measurement method
Macro	Public policy to create 41,000 miles interstate system; truck vs. rail freight haul	Defense; transportation	Congress; lobby	Congress appropriation	Cost of capital inflation	Military utilization; quality of life	% of 169.6 billion gross construction product
	40 Billion of gross construction product in 10-yr period	Labor; employment	Auto, petrochemical, and cement industries	FHWA	Cost of capital inflation; availability of resources	Need for interstate vs. federal and state systems	% of 9.6 billion transporation construction product $/ton mi frgt. haul
	Design and quality criteria	Life cycle cost; safety; economy of use; resource availability	AASTO; TRB	FHWA; state highway department	Design simplicity; degree of quality; environmental considerations	Resource utilization; esthetics	$/mile of alternate design and alternate quality
	70 mph; line of site/ grades/curves	Energy; political; safety	AASTO; A/E	State highway department; A/E	Volume of cuts and fills; no. of bridges; no. of curves	Lives saved; highway capacity	$/mile for vehicle operation; no. of lives saved
	Choice of materials; concrete vs. steel, concrete vs. asphalt	Political; life expectancy; maintenance	A/E	State highway department; A/E; contractor	Availability of particular craft labor skill	Cost histories; cost estimates	$/ft^2 of bridge
	Component cost of labor, materials, and equipment	Availability of labor and capital	A/E; contractor	A/E; contractor	Transportation of materials; storage of supplies; equipment availability	Cost histories; cost estimates	Total cost/yd^3 of concrete
Micro	Construction methods; capital vs. labor	Maximum economy	Contractor; union	Contractor	Size of crew; size of machine; work location	Capital investment per employee	ft^2 of forms; labor/equipment trade-off
	Units of work accomplished per man-hour	Minimum cost for desired quality	Contractor; union	Contractor	Weather; overtime; work conditions; changes	Work sampling	ft^2·man-hour

[a] AASTO = American Association of State Transportation Officials. TRB = Transportation Research Board. A/E = Architect/Engineer. FHWA = Federal Highway Administration.

318

alternatives, the basis of the decision, who influenced the decision, who will implement the decision, what productivity factors are involved, what measurement of cost effectiveness is used, and what measure is used to determine productivity factors at all levels. Because of the overlap between levels, the matrix is divided at the three middle levels to show greater detail.

Generally, the construction of any project represents an allocation of resources. Many decisions must be made with regard to the choice of resources and the ways they are applied. The policy decision to allocate funds to the highway sector rather than to other public projects ultimately involves a commitment to the interstate system itself considering land use, resource use, employment, and other social, economic, and environmental issues. Employment is a different kind of question at this social-planning level than at the on-site level, of course. Decisions at all the macro levels of the model involve consideration of lower-level factors.

9.6 CHALLENGES FOR THE FUTURE

Why hasn't construction productivity been the focus of an intensive, directed improvement effort? One reason is that many decision makers have failed to recognize and take into account all the elements of the productivity issue. The primary impact of the Hierarchy concept is to demonstrate that "fragmented" components of the overall construction process actually fit together in a comprehensive, predictable way. The model becomes a means to study overall productivity. In the future, increasingly stringent regulation, increasing inflation, increased union requirements, and many other factors may impinge on productivity. It is therefore essential that people at every level of the industry understand the complexities of the productivity problem. The hierarchy model is a conceptual framework for recognizing and assessing new or changing productivity factors.

9.6.1 THE ROLE OF MANAGEMENT

For the future, management will need to take a hard look at how to improve productivity. The following is a list of activities and areas where special attention is required to maintain or improve productivity.

1. Preplaning of *all* activities.
2. Scheduling that is detailed and continuous.

3. Improved purchasing practices.
4. Supervision.
5. Design (constructibility).
6. Training.
7. Productivity bargaining.
8. Prefabrication and off-site assembly where this is cost-effective.
9. Value engineering (for constructibility).
10. Study of alternative methods, processes, or materials by structured, logical means.
11. Organization of field work for most effective completion.
12. Use of man-hour targets and controls on actual labor.

Table 9.9 summarizes some of the most critical requirements for increasing construction productivity.

9.6.2 PLANNING AND DESIGN

Most of the foregoing discussion has been addressed to the building segment of the total construction cycle. Two other segments—planning and design—require increased attention in the future. Although some productivity growth is likely from work-site gains, intelligent, comprehensive planning and design are areas more likely to contribute to increased productivity.

The construction cycle begins with the planning phase. Starting with the birth of an idea and ending where the architect and engineering team begins, the planning phase generates the basic requirements of the construction project. In many cases these basic requirements present barriers to productivity that cannot be penetrated by the architects, engineers, or construction workers. To cite an example, it is a planning decision whether or not to use cut-and-cover methods for construction of a subway system. In an urban area that is heavily underlaid with buried utilities and building foundations, with a heavily traveled surface street system, such a plan creates inescapable and costly construction problems, productivity thus, plummets.

To improve the cost effectiveness of the planning activity, "constructibility" analyses should be done at the earliest possible time. Critical planning decisions are too often made by people who do not have a good understanding of the construction process. The planning problem is most severe in the public sector, where the inflow of tax money can create a "need" to build. This results in the desire to commit funds in a hurry, to designs that have poor constructibility and that will cause

TABLE 9.9 Summary of Requirements for Improving Construction Productivity

1. Communication and a more precise, common language for communication across all interfaces of the productivity problem is imperative.
2. Definition of the construction process structure in a top-to-bottom model/ framework is a key starting point.
3. Accurate identification of roles and responsibilities at all industry levels is a key factor.
4. Identification of and compensation for the diverse industry biases and interests with respect to productivity are important.
5. Development of productivity data base from the "bottom up" and dealing with all levels are essential.
6. The ability to "shape and control" operations rather than "respond and survive" needs to be developed.
7. Current contractual language is a much greater productivity detractor than many of the other factors that affect the industry.
8. Providing a larger base of trained, qualified foremen at the construction level is needed, as is a proper training of personnel at all levels.
9. Establishing and maintaining good momentum by means of forward planning, timely decision making, and labor tranquility are essential.
10. Public confidence in the industry's ability to provide a product within time and cost budgets has to be reestablished.
11. Application of traditional work sampling and analysis technology to the construction industry is an important consideration and one that could improve the analysis and cost accountability of all operations at all levels.
12. "Superprojects" offer the ability to study productivity because the large volume of similar operations provides an opportunity to perform a wide range of work sampling and analysis techniques.

construction productivity to be low. Planners trying to analyze constructibility of a proposed project must explore at least the following:

1. Cost–benefit analysis.
2. Political interests involved and potential problems.
3. Environmental impact.
4. Resource availability and allocation.
5. Long-term demographic trends of interest.
6. Construction state of the art(s).
7. Public acceptance.

Awareness is the key to success in the planning phase: awareness of the real social and corporate needs and awareness of what the construction industry can and cannot do productively. With adequate expert advice, the planning process can select alternatives that employ the most productive elements of the construction process.

Design is the one segment of the construction cycle where productivity gains can be most readily influenced. The designer, through lack of experience or carelessness, can create unnecessary expenditures. Architects and engineers frequently concentrate their efforts on the form, function, and esthetic qualities of the structure, and pay little attention to the question of constructibility. Too often designer's lack of practical construction experience and lack of understanding or interest in the process of construction make it difficult for him to identify those elements that make up a reasonably constructible project. The most important design considerations that affect constructibility are as follows:

1. Materials.
2. Modular construction.
3. Precast components.
4. All-weather construction.
5. Standardized dimensions.
6. Repetitive operations.
7. Mechanical-electrical installations.
8. Workable specifications.

A more effective design effort will greatly increase the potential for improving construction productivity. Identification of design features that are "underproductive" or difficult to construct should be made available to designers. This could be an enormous stimulus for improving the rate of productivity growth in the industry.

9.7 SUMMARY

The discussion in this chapter provides a perspective on construction productivity and of the inherent complexities involved in understanding or improving productivity. The following summary of the material states in practical terms some of the problems or issues of construction productivity. A number o points made previously in the chapter are restated with this emphasis.

Efficiency and effective work are major concerns.   Increased productivity results not from working harder but from working smarter. Here the key is to study production methods and find those that enable workers to work effectively.

Size of job and number of workers is a productivity factor.   In some cases, 2000 workers or so was found to be a break point in job productivity. Getting everyone on a job working toward one goal gets very complicated when there are too many people involved.

Productivity improvement methods may not be transferable from one project to another.   Be cautious in attempting to generalize from quantitative analyses of the work on one project. Qualitative comparisons of productivity factors can be transferred more readily from one job to another and between types of construction work.

Productivity improvement begins long before the on-site work.   It must be a goal of the entire process, starting at the earliest planning phases.

Owners have an important role.   The owner can force the productivity issue through contract requirements and/or administrative procedures such as (1) detailed scheduling techniques; (2) documentation for progress payments; and (3) productivity measurement and control systems. Owners benefit by the use of these systems even if they must pay for them.

Timeliness of decisions and good communication between the various levels of the hierarchy contribute to high productivity.   Communication between levels is often missing entirely. Decisions frequently must go all the way to the topmost levels to be resolved; the more levels involved, the longer it takes to get the decision made.

Management style is an important factor.

The superintendent is still the boss.   Improvements in on-the-job productivity are greatest when the superintendent is included in planning for change. Programs to increase productivity will be most successful with his help and participation and that of the crew foremen.

Quality   assurance   and   quality   control   are   important concerns.   Consistent quality throughout the job raises the value of the product. This constitutes improved productivity (increased value produced per unit of input).

Gross productivity cannot be directly compared between specialties.   Some kinds of work are normally less productive; likewise, sometimes workers must wait for the completion of someone else's work before they can start. This makes them less productive. It is essential to keep in mind that productivity, even on similar jobs, can vary 500% even if the work is done relatively efficiently. For instance, where work is paced (e.g., siding, brickwork, steel, concrete, insulation of boilers), productivity is high. When the same work is done as fill-in work, productivity will be low but the work is still effective.

Individual workers are concerned with the quality of working life. These concerns affect productivity in the long run. If workers feel management is concerned only with numbers and dollars, and not with the work or the crews, motivation will be low. Examples of the concerns that make up that intangible quality of the work environment are safety conditions, opportunities for breaks, pleasant conditions when possible, and so on.

Construction workers generally want to work effectively and to feel productive. To intelligently employ manpower, management must establish a planned work environment for effective operation, considering all the related work in progress. Restrictive work practices and other labor requirements can often be handled without reducing productivity. Planning and understanding of the work process is required to employ labor most effectively.

Improvement of labor productivity does not mean a general loss of jobs on the labor market. The increased ability to employ manpower more cost effectively means that labor-intensive construction methods will tend to maintain their competitive advantage over machine-intensive alternatives. On a broader economic level, increased productivity of the construction industry itself means that construction of new facilities will have a competitive edge over other methods of producing profits in the private sector and social benefits in the public sector.

REFERENCES

Bureau of Labor Statistics, U.S. Department of Labor. *Improving Productivity: Labor and Management Approaches.* Washington, D.C.: U.S. Government Printing Office, September 1971.

Cassimatis, P. J. *Economics of the Construction Industry* (Studies in Business Economics No. III). New York: National Industrial Conference Board, 1969.

Construction Industry Research. *Listing of Construction Industry Cost Effectiveness References.* New York: Business Roundtable, February 1980.

Crandall, K. C. *Productivity Research for the Construction Industry.* Final Report on Workshop for the Formulation of Specific Projects (Tech. Rep. 5). Berkley, Calif.: Civil Engineering Department, University of California, 1978.

Hinze, J. and Parker, H. Safety: productivity and job pressure, *Journal of the Construction Division, American Society of Civil Engineers,* **104,** 27–34, March 1978.

Kellogg, J. C., Howell, G. E., and Taylor, D. C. *The Hierarchy Model of the Construction Industry*. Report for National Center for Productivity and Quality of the Working Life. Littleton, Colo.: Kellogg Corporation, April 1978.

Parker, H. W., and Oglesby, C. H. *Methods Improvement for Construction Managers*, New York: McGraw-Hill, 1972.

Samelson, N. M. *Effect of Foreman on Safety in Construction* (Tech. Rep. No. 219). Palo Alto, Calif.: Stanford University, Department of Civil Engineering, 1977.

Sutermeister, R. A. *People and Productivity*, 3rd ed. New York: McGraw-Hill, 1976.

Taylor, D. C., Kellogg, J. C., and Wilkinson, M. C. A construction industry R&D incentives program. *Engineering Issues—Journal of Professional Activities, ASCE*, **102**, (E13),369–390. Proc. Paper 12271, July 1976.

Taylor, D. C., *Bibliography on Construction Productivity and Measurement*. Prepared for National Center on Productivity and Quality of Working Life. Littleton, Colo.: Kellogg Corporation, April 1978.

Thomas, C. W., Albertson, M. L., Taylor, D. C., Wisely, W. H., and Baker, R. F., *Research Needs in Civil Engineering Relevant to the Goals of Society*. Prepared jointly by Colorado State University and American Society of Civil Engineers. New York: ASCE, June 1971.

U.S. Department of Commerce. Expenditures: Construction Industry. *Industry and Trade Administration*, **25**, 10, November 1979.

CHAPTER 10

WOMEN AND MINORITIES IN CONSTRUCTION: THE IMPACT OF AFFIRMATIVE ACTION AND ITS EFFECTS ON WORK PRODUCTIVITY

CONSTANTINA SAFILIOS-ROTHSCHILD, Ph.D.

Department of Human Development
Pennsylvania State University
State College, Pennsylvania

10.1 WOMEN IN CONSTRUCTION

Although the positions of women and minority men in construction have largely followed similar patterns, there are sufficient differences to warrant their separate treatment. Both groups have been kept away from well-paid jobs in developed societies, but the stereotypes used to exclude them and the success of corrective mechanisms to bring them into construction have been quite different.

10.1.1 CROSS-CULTURAL AND HISTORICAL TRENDS

A cross-cultural and historical perspective on women's access to different occupations usually follows a typical pattern: as long as the occupation is a poorly paid and low-prestige one, or as long as it is informal and not yet organized as an occupation, women have access to it. However, as soon as the occupation becomes formally organized and/or upgraded in terms of pay or prestige, men enter the occupation, displacing women. These women are pushed into less well-paid, less prestigious occupations (Safilios-Rothschild, 1974). The reverse trend also holds true: as long as an occupation such as secretary or servant is relatively well-paid and desirable (as is the case in India, Pakistan, and many African countries), men compete with women and win these jobs.

Construction work fits the pattern described above. In many developing countries, in which the construction of houses and other buildings is carried out informally through a variety of unpaid self-help schemes, women are actively involved in it. For example, it is reported that in Africa, often more than 50% of the work involved in the construction of roads, nursery schools, primary schools, and village centers is women's work. More specifically, women provide 80% of the self-help construction labor in Kenya under Food-for-Work and other programs (*The Role of Women in African Development*, 1975). According to other calculations, 30% of the labor needed in Africa for house building and 50% of the labor needed for house repair is contributed by women (The Changing and Contemporary Role of Women in African Development, 1974). Similarly, Indian women, especially those in urban areas, participate actively and widely in the building trade (Boserup, 1976). Even in developing societies such as Egypt where, at present, women do not work in construction, historical accounts show that low-income women did construction work in the nineteenth century.

In fact, worldwide statistics show that construction work not only represents a "traditional" form of work for women in many third world countries, but also that it is not an unusual occupation in eastern European societies such as Hungary, Poland, Czechoslovakia, Bulgaria, Yugoslavia, and Romania. The same is true in Hong Kong, Japan, Finland, Austria, and Monaco (Boulding et al., 1976). These last-named examples demonstrate that women in construction can be found in developed societies in which women enjoy a wide range of occupational options, as well as in developing countries where women build their own houses and community buildings (or are found as unskilled construction workers in cities).

Why have women in the United States and other highly industrialized societies been almost excluded from construction work? The main reason is, of course, the fact that construction work is organized, unionized, and highly paid, and men have had a high stake in gaining and maintaining exclusive control over access to it. In this way not only women have been excluded from the unions and the opportunity to work in construction, but also blacks and other ethnic groups. A variety of stereotypes have helped keep women out of construction work, stereotypes that have little to do with the work reality of women in construction in the Third World, eastern Europe, Finland, or Austria.

10.1.2 STEREOTYPES JUSTIFYING WOMEN'S EXCLUSION FROM CONSTRUCTION

Despite the fact that in the 1950s and early 1960s in the United States lower-middle, working, and lower class women had to work much more often than middle class women, they seldom held well-paid, skilled jobs. They worked in "feminine"-stereotyped jobs with a high turnover, short training requirements, and low pay. In the late 1960s, however, mainly as a result of a changing climate regarding the roles of women, more women began to be employed in skilled trades even before affirmative action was begun. Table 10.1 draws from the 1970 census and shows the changes that took place in the employment of women in construction-related skilled trades. These data show that, although between 1960 and 1970 the percentage of women in the different construction trades tripled or more, still the percentages were extremely low (except among paperhangers, 10.8% of whom were women). Women

TABLE 10.1 Percentages of Women in Skilled Construction Trades, 1960 and 1970

Trade	1960	1970
Electricians	0.7	1.8
Plumbers and pipefitters	0.3	1.1
Sheet-metal workers	1.1	1.9
Bricklayers	0.5	1.3
Machinists	1.3	3.1
Boilermakers	0.2	1.3
Plasterers	0.3	1.5
Carpenters	0.4	1.3
Glaziers	1.3	3.1
Painters	1.9	4.1
Paperhangers	6.0	10.8
Roofers and cement finishers	0.2	1.3
Supervisors/construction	0.2	1.1
Inspectors/construction	0.7	1.5

became attracted to construction jobs mainly because of the good pay, but also in some cases because their fathers or other male members of their families were in construction and were familiar with the advantages and possibilities in the occupation (Gizyn, 1973).

However, although the 1970 (and 1973) Bureau of Labor Statistics figures show that 6% of those employed in construction were women, the vast majority of them still held white collar, "women's" jobs, working in the offices of contractors, subcontractors, and architects, for example (Gizyn, 1973, 1974). Also, the data in Table 10.2 show that the two categories of building trades in which women were better represented in 1970, paperhangers and painters, were also the worst-paid trades. Thus, in general, the trend has been the more skilled a trade (such as electrician, plumber, carpenter), the more it was labeled "masculine" and declared off-limits to women. Moreover, in 1972, women were even more poorly represented in the membership of local referral building trade unions among 16 international such unions of

TABLE 10.2   Average National Hourly
Wage for Crafts in the Building Trades
1972

Trade	Average hourly wage
Electricians	8.19
Plumbers	8.15
Pipe fitters	8.14
Sheet-metal workers	8.09
Asbestos workers	8.01
Elevator constructors	8.00
Bricklayers	7.99
Ironworkers	7.79
Lathers	7.67
Boilermakers	7.59
Plasterers	7.45
Carpenters	7.41
Roofers, composition	7.37
Cement finishers	7.24
Roofers, slate and tile	7.22
Paperhangers	7.09
Painters	7.06
Laborers	5.68

SOURCE: U.S. Department of Labor, *BLS Bull. 1807.*

TABLE 10.3 Minority and Female Members as a
Percentage of Total Members, International, 1972

International	Minorities (% of Total)	Women (% of Total)
Total	15.6	0.7
Electrical workers, I.B.E.W.	7.5	2.3
Laborers	43.4	1.0
Painters and allied trades	14.9	0.6
Carpenters	11.2	0.5
Operating engineers	6.2	0.5
Sheet-metalworkers	7.0	0.4
Boilermakers	11.3	0.2
Plumbers and pipe fitters	4.4	a
Ironworkers	9.3	a
Plasterers and cement workers	32.6	a
Bricklayers	13.1	a
Roofers	23.3	a
Elevator constructors	5.6	0.0
Lathers	14.2	0.0
Marble polishers	15.2	0.0
Asbestos workers	3.7	0.0

SOURCE: 1972 EEO-3 Reports.
[a] Less than 0.05%.

key trades in construction (see Table 10.3). Their membership was in fact many times lower than that of minority groups, especially in trades such as plumbers and pipe fitters, ironworkers, plasterers and cement workers, bricklayers, roofers, elevator constructors, lathers, marble polishers, and asbestos workers, where there were hardly any women. This low membership resulted in women not being referred to employers seeking to hire workers in these trades. Thus even women with the appropriate skills were not employed (*Minorities and Women in Referral Units in Building Trade Unions, 1972, 1974*).

The exclusion of women was rationalized on the basis of a range of stereotypes about women's temperament, physical strength, and ability as well as behavior. These stereotypes partly reflected protective attitudes toward the biologically "weaker" female as well as beliefs that women cannot be reliable, serious workers because of their interfering familial responsibilities. This was the essence of sex discrimination in the occupational world.

A 1970 Wisconsin survey study of journeymen and supervisors with or without women skilled workers showed that two-thirds of the respondents would hesitate to have women apprentices in some trades which are not "suitable" for women because they involve long hours or are too "dirty" or "heavy." More than a quarter of employers felt that whereas some trades such as sewing, upholstery, interior decorating, and drafting require precision and manual dexterity and are well suited for women, others are not (*Women in Apprenticeships— Why Not?*, 1974).

Despite these stereotyped beliefs, in about one-third of the shops women were in fact employed in "unsuitable" skilled trades which required mechanical aptitude, were dirty, or were heavy. This fact was not widely known by the men in the surveyed shops, probably because the fact went counter to their beliefs (*Women in Apprenticeships—Why Not?*, 1974). Thus the "protective" stereotypes did not consistently keep women out of skilled jobs partly because some male managers did not adhere to them. But the fact that these women could do the jobs, even when known, tended to be written off as representing the performance of "unusual" women and hence was not to be generalized to other interested women. Furthermore, research on sex differences in terms of aptitudes regarding skilled trades have consistently shown either no sex differences or that women excel in more aptitudes than men (Hedges and Bernis, 1974).

In addition, since at least the skilled trades in construction require a lengthy apprenticeship, a number of structural and internal barriers were (and still are) keeping women out. The structural barriers were the fact that high school vocational counselors actively discouraged girls or simply did not recommend them for apprenticeship openings. In most schools girls were overtly forbidden or subtly discouraged from taking shop courses or developing an "unfeminine" technical competence (*Women in Apprenticeships—Why Not?*, 1974).

Age limits have been another important barrier. Construction Trade Joint Apprenticeship Committees, for example, usually had set an upper age limit between 24 and 27, which excluded the majority of interested women, who were in their late twenties or thirties. Younger women accepted the idea that working in a "male" occupation would make them unfeminine and therefore undesirable to men. Besides, the long apprenticeship was not attractive to young women in the 1960s who saw their life plans primarily or entirely in terms of marriage and motherhood. The women in their late twenties or early thirties who became interested in apprenticeships had already experienced the realities of the job market (and often of their private lives as well) and were attracted to good working opportunities in order to support themselves and their children (*Women in Apprenticeships—Why Not?*, 1974).

10.1.3 AFFIRMATIVE ACTION AND ACCESS TO CONSTRUCTION JOBS FOR WOMEN

Title VII of the Civil Rights Act of 1964 provided the framework to assure equal occupational opportunity for women, and Executive Order 11375 in 1968 added sex discrimination to the other types of discrimination prohibited by the 1965 Executive Order 11246. These executive orders, however, covered only contractors with $10,000 or more in federal contracts or other contracts and included colleges and universities but did not bind unions and exempted state and local governments. Enforcement of the order was vested in the Office of Federal Contract Compliance (OFCC), an agency of the Department of Labor.

Initially, there was a controversy over the term "quotas," which was criticized as promoting the hiring of unqualified persons. Instead of quotas, plans and goals were required by the Department of Labor. However, the implementation of these requirements for affirmative action plans has been more effective and systematic with regard to minorities than with regard to women. The many "hometown" affirmative action plans for different cities required nondiscrimination on the basis of sex but specified goals and timetables only for minorities, not for women (*Construction: The Industry and the Labor Force*, 1976). The result of this rather lax attitude toward the employment of women was no significant change in the percentages of women working in construction-related skilled trades which still, in 1976, ranged between less than 1 and 3% of the total. Although the number of women apprentices between 1974 and 1975 increased by 74%, only 1.2% of all apprentices were women (*Equal Employment Opportunity in Apprenticeship and Training*, 1978). Government agencies and training programs often ignored women. Thus Manpower Training Programs did not give women the opportunity to learn through the on-the-job apprenticeship ladder, and the Labor Education Advanced Program, specializing in apprenticeship placements for minorities, ignored women. On the other hand, the coordinators of the Work Incentive Program (WIN) lacked knowledge of apprenticeships and the skilled trades. Therefore, systematic efforts to increase the numbers of women apprentices did not begin until the mid-1970s when the Urban League, Recruitment and Training Programs, the Human Resources Development Institute, and Opportunities Industrialization Centers, often jointly with the YWCA, NOW, and the League of Women Voters, initiated apprenticeship outreach programs for women. At present, women can get information, assistance, and training in construction trades in most major and some smaller cities in the United States, but these newly available services and opportunities are only beginning to make an impact (Rich, 1978).

The failure to reach out to women and include them in apprenticeship programs for construction trades facilitated employers' tendency not to hire women for construction jobs. Employers could thus hide their reluctance to hire women under the convenient guise of "not enough trained women," and construction jobs continued to be inaccessible to women.

It was only in 1978 that the Federal Government responded to the unsuccessful, passive nondiscrimination-based-on-sex stance taken by the construction industry. Specific hiring guidelines went into effect on May 7, 1978, according to which 3.1% of the construction work force must be women in 1978, 5% in 1979, and 6.5% in 1980, and hometown plan administrators were required to submit detailed documentation as to how they intend to achieve these goals (Labor Department Holds to Rigid Rules for Construction Hiring Plans, 1978). These guidelines were enforced, as evidenced by the fact that two voluntary hometown minority hiring plans in Buffalo and Rochester were not renewed in September, 1978 because they did not submit detailed plans on how they would implement the new Federal goals and timetables for hiring women (Labor Department Kills Two Hometown Plans, 1978).

While the hiring guidelines for women were being firmed, in June of 1978 the Department of Labor also issued regulations on Equal Employment Opportunity in Apprenticeship and Training. According to these regulations, since women are now 41% of the nation's labor force, the goals for entry apprenticeship classes should not be less than 50% of that figure. Hence in most regions, beginning apprenticeship classes must aim for 20% women, to be achieved through outreach and positive recruitment. In addition, goals and timetables must be updated annually until there are as many women apprentices as there are women in the work force, and minimum physical requirements often representing barriers to women must be eliminated (*Equal Employment Opportunity in Apprenticeship and Training*, 1978).

By the end of 1978, the goal of 3.1% of the construction work force being women was not achieved, partly because the apprenticeship system was not yet producing enough women to meet the federal hiring goals (A Record Apprenticeship Crop, 1978). Increasingly, however, the schemes and initiatives to train women in building trades are multiplying. The construction industry has begun to respond and federal grants facilitate the process. For example, federal grants by the Department of Labor are used to train welfare women in building trades in an 11-week summer program at the Indiana Laborer's Training Institute established by the state's construction industry (Shulins, 1979).

The pressure to train and hire more women in building trades brought a controversial response from the Associated General Contractors (AGC), namely, a 2000-hr (1-year) training program for women, instead of the traditional 3300-hr apprenticeship training. This shorter training program is defended by AGC as not being proposed as a substitute for apprenticeship, but it is, in fact, fought by the Labor Department's Employment and Training Administration and the AFL-CIO Building and Construction Trades Department. The YWCA and the National Association of Women in Construction have supported it, however, because it provides women with a flexible, valuable alternative to apprenticeship. Of course there is a danger in the institutionalization of such a shorter program to be used *only for women* in that it tends to create a "lesser" category of women skilled in construction trades which can be used to justify lower pay and lower status for women. Furthermore, this lesser training may block women's entry into the unions.

10.1.4 THE IMPACT OF WOMEN ON WORK PERFORMANCE

The crucial questions in everybody's minds are, how do women work out in jobs previously performed only by men? Do they create any problems? How do the men around them feel? Does work productivity suffer? The answers to all these important questions are not simple. Radical changes always bring about inevitable, painful transition problems. Let us look at those problems as well as their possible solutions.

First, the fact that at present women have entered the construction trades in very small numbers creates a series of problems that are work-related and have nothing to do with the federally imposed guidelines. The very small number of women carpenters or women pipe fitters, for example, makes them *very visible*, so that behavior is constantly scrutinized and evaluated by the men around. At the beginning, the women are tested often by being assigned the toughest tasks to see whether they can "make it." When they do not perform satisfactorily, their poor performance is generalized to all women. When they perform well, their performance is written off as unusual (Safilios-Rothschild, 1978). This is a "no win" situation.

Second, as long as there are only a few women in construction trades, the men around them have serious difficulties relating to them as co-workers rather than as potential lovers. The effect of the small number of women is accentuated by the newness of the situation. The men almost feel that they have to try to take them out, to turn them into lovers. Therefore, women are often sexually harassed. At best, women are at first a distraction, a new challenge with which the men must learn to cope (Rich, 1978; Riemer, 1978). And it must be quickly added that not all women remain passive in this situation. Some women go out with their supervisors because they think it is the "smart" thing to do in order to get ahead. Others like to flirt and be able to choose among many men. There are many who are ambivalent and send contradictory messages to the men around them. In fact, women are not better equipped than men to deal with this new type of work situation, and it is possible that the work performance of all is initially affected. The only real solution to this problem is to increase the numbers of women in order to slowly "normalize" the interpersonal relations between men and women workers by rendering the presence of women routine and expected. Only in this way can men in construction and other previously all-male occupations overcome their nostalgia for the "all-male-club" atmosphere and stop resisting employment of women.

Of course, it must be stated here that in some cases, men's resistance to the presence of women is much more clear-cut and violent. Some men do everything they can to cause the women to fail; they try to break down their self-confidence and insult and frustrate them until they see them cry, or, more rarely, they are actually violent toward the women (Rich, 1978).

Third, the fact remains that women, on the average, have less physical strength than men and that some of the construction jobs and tools tax women's physical endurance more than that of men. We know, however, that men do not like to admit that a job is breaking their backs or that the use of a particular tool creates problems for them, because such admissions are not compatible with the "macho" image of a construction worker. The first time, therefore, that a man with a work-related problem is seen and heard is when he is hurt and when he usually has cost a lot of money to the company that employs him in terms of different work placement or workmen's compensation.

Feminist ideology has often clouded this issue since it has been considered politically expedient, in order to stop the eternal biological argument from constraining women's options, to argue that few construction jobs still require physical strength beyond women's capacities. There is, however, a more efficient and appropriate solution, namely, the application of human factors research to the restructuring of "heavy, hard" jobs as well as to design of tools. The needed adjustment of tools is often related to the fact that, on the average, women's hands are smaller than men's and therefore women have trouble with grips of tools made for men (Ducharne, 1975). It is interesting to note that as women are struggling to be accepted as equals with men in "male" jobs, they rarely dare complain about the difficulties encountered with the tools used. Instead, they often wind up as medical cases, replicating men's behavior.

The needed adjustments are often simple (for example, changing the grip of welding guns) and relatively inexpensive. But even when they are expensive, the expense is well worth it for a number of reasons. In addition to providing women with more employment options and to cutting down the injury and disability rate for both men and women, they in fact end up saving money to the companies which undertake them by cutting down the costs for absenteeism, medical care, rehabilitation, and workmen's compensation.

Finally, as women become a familiar sight in construction, some employers begin to discover distinct advantages in women's work performance. The personnel director of a North Carolina construction company whose work force ranges from 10 to 33% women reported in a workshop that women are better workers than men because they do not "hot rod" while operating heavy equipment, they are not out drunk on Monday, and they keep a clean and well-organized warehouse and a good inventory. Although these comments are encouraging regarding women's acceptance in construction, it is striking that the praises follow sex role stereotypical beliefs about women's neatness, housekeeping skills, and moral, obedient, and orderly behavior—stereotypes that may no longer fit the majority of women (especially young women).

10.1.5   WOMEN AS INDEPENDENT CONTRACTORS

As a response to the prevailing discrimination against women and to the existing barriers to women, women have become independent contractors and managers in construction. This development reflects a broader, national trend according to which women have been entering more and more small business undertakings. In some cases, this development follows traditional lines in that the women take over businesses begun by their fathers or husbands. Often the women were helping their husbands (she was his "right hand") and had only to move from an auxiliary to the principal position (Gizyn, 1973, 1974;

Little Blond Widow Runs Thriving Construction Firm, 1978). Less often, women come up the ladder from secretarial positions after having received the appropriate technical and/or business training.

In some cases, sex role stereotypes seem to help these women succeed in their businesses. The women contractors, who seem to share with their customers these stereotypical beliefs, feel that their customers appreciate the advice on remodeling (e.g., color for fixtures and level of noise of waste disposers) they can get from them, although the remodeling is a small fraction of the main job of plumbing repair (Gizyn, 1976). Other women contractors feel that their presence renders the negotiation climate calmer and "more conducive to getting things done" (Gizyn, 1973). The personal accounts of some of these women contractors in construction suggest that the few women in this all-male undertaking are eager to "soften" their intelligence, competence, knowledge, and skills by projecting a very "feminine" image ("Little Blond Widow Runs Thriving Construction Firm," 1978). Also as a response to discrimination against women on the part of construction trade unions, women in these trades have organized their own National Association of Women in Construction (NAWIC) which, in 1976, had 7120 members.

Women have not come a long way in building trades and certainly have a long way to go.

10.2 MINORITIES IN CONSTRUCTION: UNDERUTILIZATION, PREJUDICE, AND COMPETITION

The pattern of underutilization of minorities in some building trades has been quite different from the systematic exclusion of women from all construction jobs. (It must be clarified here that by minorities, consistent with the way the term has been used in the construction-related literature, we refer to minority men rather than minority women. Minority women have been systematically excluded from construction work along with white women.) Minority men were never entirely excluded from construction; they were selectively utilized in the heaviest and least paid jobs whenever the supply of white manpower was not sufficient (or in cases of strikes). The greatest concentration of black men, for example, has been among construction laborers who perform the heaviest, the least skilled, and the lowest paid jobs in construction (Landon and Peirce, 1972). In fact, the distribution of minorities in the different building trades varies inversely with the degree of "desirability" of the trade in terms of skill, wage rates, and extent of individual union involvement. The most prestigious electromechanical trades, including electricians, plumbers, pipe fitters, iron and sheet-metal workers, asbestos workers, and elevator constructors, have had the lowest percentages of minorities (see Tables 10.4 and 10.5). The percentages of minorities have been progressively higher from "intermediate" trades such as bricklayers, plasterers and cement masons, roofers, paperhangers, and laborers (Hammerman, 1972; see also Table 10.3).

Entry to the highly skilled trades and, relatively less so, those of intermediate skill has been regulated by the unions, and minorities have been effectively kept in small numbers through a number of processes and mechanisms that kept them out of apprenticeships as well as union membership. Apprenticeship is more essential for highly skilled trades than for other building trades in which minority as well as white men could gradually gain the necessary skills and experience on the job (Foster and Strauss, 1972). Furthermore, minority men could always be found in greater percentages in nonunion construction jobs which were paid less and required less formal training experience such as residential construction jobs (Foster, 1974).

The processes and mechanisms that kept minority men out of apprenticeships were largely the same as those that kept them out of trade unions, namely, outright discrimination against minorities; preferential treatment of sons, people of the same ethnic origin, and political allies; secretive recruitment and selection processes of which only a few people (and certainly not the minorities) are informed; failure of minorities to aspire to becomes apprentices; and finally, entrance requirements such as high school graduation or ability to pass an admissions examination (Foster and Strauss, 1972; Dubinsky, 1971). Probably in some cases the fear of competition on the part of minorities strengthened by the fear that minorities may play a scab/strikebreaking role, as has been true in the past, may be as influential as outright prejudice against blacks and other minorities (Dubinsky, 1971). The influence of racial and ethnic stereotypes must not be played down, however, especially since these stereotypes have been quite potent and negative. Blacks (and chicanos) have been seen as lazy, irresponsible, unreliable, of low intelligence and ability, poor workers, quarreling, drinking, and taking time out of work. They were therefore thought to be fit for heavy jobs in which strong muscles and little ability and responsibility were required (such as laborers) and in which "bodies" rather than individuals are important and, hence, easily replaceable.

TABLE 10.4 Distribution of Total and Minority Referral Unit on Membership in the Building Trades, 1972

International Total	All members	Minority group members			
		Black	Spanish surname	Asian American	American Indian
Number	1,604,451	133,572	96,940	5,828	14,439
Percent	100.0	100.0	100.0	100.0	100.0
Asbestos workers	0.6	0.1	0.1	0.1	0.5
Boilermakers	2.0	1.1	1.7	2.2	3.0
Bricklayers	2.0	2.4	1.0	0.1	1.0
Carpenters	22.8	10.1	22.7	53.3	17.0
Electrical workers, I.B.E.W.	14.8	4.6	9.9	11.3	9.6
Elevator constructors	0.6	0.2	0.2	0.2	0.2
Ironworkers	5.3	1.5	3.6	5.3	14.3
Laborers	18.4	64.4	39.4	14.6	23.4
Lathers	0.2	0.1	0.3	0.2	0.2
Marble polishers	0.2	0.1	0.4	a	0.1
Operating engineers	11.4	5.3	2.3	1.9	12.7
Painters and allied trades	4.2	2.5	6.3	4.6	2.8
Plasterers and cement workers	1.8	3.5	4.6	0.8	2.0
Plumbers and Pipe fitters	11.8	2.1	4.1	4.2	9.4
Roofers	0.9	1.6	1.0	0.2	2.3
Sheet-metal workers	2.9	0.4	2.5	1.1	1.4

SOURCE: 1972 EEO-3 Reports.
a Less than 0.05%.

TABLE 10.5   Percentage of Black Union Membership in Higher-
Paying Electromechanical Trades 1967–1972

Year	Electricians	Ironworkers	Plumbers and pipe fitters	Sheet-metal workers
1967	1.6	1.7	0.2	0.2
1969	1.9	1.7	0.8	0.7
1971	1.8	2.2	1.2	1.0
1972	2.6	2.4	1.5	1.1

SOURCE: EEOC Office of Research.

10.2.1   AFFIRMATIVE ACTION: CHANGES AND RESISTANCE

Affirmative action for minorities and especially for blacks has had a
considerable head start over affirmative action for women. Title VII
of the Civil Rights Act of 1964 and the Executive Order 11246 in 1965
prohibited racial, religious, and ethnic discrimination. And already in
October 1971, the regulations of the U.S. Department of Labor Bureau
of Apprenticeship and Training (BAT) required goals and timetables
relative to equal employment opportunity for minority workers for
continued approval of apprenticeship and training programs. Therefore, these goals and timetables in construction were required by the
federal government 7 years earlier than for women, a significant headstart. The important questions, however, are as follows: did this headstart make any difference? Have minorities' access to construction jobs,
especially highly skilled and highly paid ones, improved more than
has been true for women? And besides the headstart of the affirmative
action for minorities, are there any crucial, substantive differences in
the implementation style that might account for the differential success of the two types of affirmative actions in construction?

   Before these questions can be answered, it is important to note that
the different regulations relating to equal employment opportunity
and to the required steps to be taken by the construction industry to
redress the existing inequalities often have been confusing, ambigu-
ous, or even contradictory. Thus they were baffling and easier to cir-
cumvent as well as difficult to effectively implement (Greuenberg,
1971). The "imposed" Philadelphia plan originally submitted in 1967
and its subsequent revisions and amendments represent a good example of how, within a period of 4 years, the specification of numerical
goals in needed minorities in each building trade was first declared
illegal, then legal for trades in which minorities were grossly unde-

rutilized, and then legitimate for all trades (Glover and Marshall, 1977). Furthermore, it is difficult to assess just how much progress has been made in making construction jobs accessible to minorities. Except for apprenticeship enrollments, there are no systematic, longitudinal data available that would allow us to evaluate. It seems that the Equal Employment Opportunity Commission (EEOC) and other governmental agencies have failed to gather the necessary data base. The requirement, for example, that construction unions submit regular reports on their minority membership has not been enforced, and the available data are not appropriately analyzed to allow longitudinal comparisons (Foster, 1978). Besides, the EEOC survey is limited to referral unions with 100 or more members and apprenticeship programs with 25 or more participants, so that it covers only an estimated one-third of the construction labor force (Glover and Marshall, 1977). Finally, it should be noted that the economic depression of 1972–1973 considerably interfered with the beginning positive results of affirmative action.

With these limitations in mind, let us see how well minority men have done in gaining greater access to the construction industry. With regard to apprenticeships in building trades, there is considerable evidence that minority men (especially black men) have made significant gains. Exactly how much gain is rather difficult to ascertain (Foster, 1978) but the following give some indications. Data on programs served by the Bureau of Apprenticeship and Training show that in 1958, minorities accounted for 7.2% of construction apprentices, and in 1972, 15.1% (Glover and Marshall, 1977). On the other hand, data recorded by the U.S. Department of Labor between January 1 and June 30, 1975 indicate that 18.4% of the apprentices were minority. (State and National Apprenticeship Program System (SNAPS) Report, 1975). Table 10.6 presents the detailed apprenticeship trends indicating that, at least in terms of enrollments, minorities were better represented even in the electromechanical construction trades. Completion data, however, although not systematically available, are less satisfactory. It is also important to note that Apprenticeship Outreach Programs have been effective in increasing the number and percentage of minorities in skilled construction crafts at a relatively low cost. In 1974, they accounted for 40% of all new minority apprenticeships in 15 major trades, this percentage being particularly high in the mechanical trades (Glover and Marshall, 1977).

With regard to union membership in the building trades, fewer gains can be documented since the unions are often mentioned as the stumbling block to the integration of minorities in construction (Jackson and Fossum, 1976). There is, however, considerable difference between the policy orientation of national union leaders and local union leaders since the former are much more sensitive to wider political issues, are concerned with their national image and status, and are interested in increasing their membership. Local leaders, on the other hand, most often wish to maintain control and to restrict membership in order to protect wage rates (Glover and Marshall, 1977; and Marshall et al., 1975). Despite these union barriers, one study based on interviews with 1234 journeymen in 28 local unions among six construction trades showed that 9% of the surveyed journeymen were minorities and that the percentage of minority union entrants from 1960 to 1972 was double that prior to 1960. But even more important, the same study showed that having completed an apprenticeship program was not a requirement for admission to a trade union. In fact, during the period 1960–1970, twice as many minorities entered the trades through non-apprenticeship routes (Glover and Marshall, 1977).

Furthermore, there is some evidence that minorities often, but not uniformly, may gain access to construction jobs to a greater extent in residential than in commercial construction mainly because it is non-union, it has to depend on locally available workers, and it requires an occupational mix that favors blacks. Most of the needed workers are carpenters, laborers, cement finishers, and sheetrock installers, that is, trades in which there are considerable numbers of blacks (Foster, 1974). Federal regulations have not touched residential construction, which constitutes over 40% of all construction, but city-wide cooperation on affirmative action plans could render the atmosphere more conducive to minority utilization. It seems, therefore, that apprenticeships and union membership are by no means the only two areas to concentrate on in affirmative action in order to expand minority utilization in the construction industry.

With regard to town-wide affirmative action programs, "imposed" and "hometown" affirmative action plans have had mixed results from town to town, but in general, "hometown" plans have been more successful to some extent than "imposed" plans. In "imposed" plans, federal or federal-assisted contracts over $500,000 "must agree to meet established minority employment goals for specified crafts throughout the life of the contract" (Rowan and Brudno, 1972). Under these plans, the focus is on hiring black workers even if only for the life of the project, thus providing them with needed work experience and, it is hoped, making the construction labor market more willing in the long run to hire other black workers (Rowan and Brudno, 1972). Voluntary "hometown" plans, on the other hand, involve unions, all local con-

TABLE 10.6 Percentage of Minorities in Apprenticeship, January 1–December 31, 1977[a]

Type of Apprenticeship	%
Air conditioning and refrigeration mechanics	21.1
Aircraft mechanics	9.3
Auto and related mechanics	21.7
Auto and related body repairs	19.3
Barbers and beauticians	18.9
Boilermakers	13.9
Bookbinders and bindery workers	15.7
Bricklayers, stone and tile setters	22.8
Butchers and meat cutters	24.3
Cabinetmakers, wood machinists	21.9
Car repairers	17.0
Carpenters	18.5
Cement masons	46.7
Compositers	11.0
Cooks and bakers	28.1
Drafters	10.5
Electrical workers	15.8
Electricians	15.8
Electronic technicians	20.6
Floor coverers	21.4
Glaziers	21.9
Industrial technicians	26.3
Insulation workers	23.1
Lathers	28.4
Line Erectors light and power	13.6
Lithographers photoengravers	15.2
Machine Setup and operators	20.8
Machinists	14.1
Maintenance mechanics	15.9
Mechanics and repairers	15.7
Medical and Dental technicians	17.3
Millwrights	15.6
Molders and coremakers	20.9
Office machine servicers	17.5
Operating engineers	31.4
Optical workers	12.1
Ornamental ironworkers	27.9
Painters	25.9
Patternmakers	6.2
Pipe fitters	17.7
Pipe fitters–steam fitters	35.2

TABLE 10.6 (*Continued*)

Type of Apprenticeship	%
Plasterers	38.4
Plumbers	14.6
Press operators	15.9
Printing and publishing workers	11.7
Radio and TV repairers	12.6
Roofers	29.7
Sheet-metal workers	18.9
Sprinkler fitters	16.7
Stationary engineers	17.2
Structural steel workers	21.1
Tapers and dry-wall installers	23.1
Toolmakers and diemakers	7.7
Miscellaneous trades	20.1
U.S. total	18.4

[a] U.S. Department of Labor, Employment and Training Administration, State and National Apprenticeship System (SNAPS).

struction contractors, and *ad hoc* coalitions of minority community organizations negotiating their own area-wide goals subject to approval by the Department of Labor's Office of Federal Contract Compliance Programs (OFCCP). Voluntary plans focus not only on the hiring of minorities but also on their becoming full members of the unions and being able to work beyond any given contract (Rowan and Brudno, 1972). In 1977, eight cities had imposed plans and 63 localities (including some states) had voluntary, "hometown" plans, but about one-third of them have since been disapproved (Foster, 1978).

Both types of affirmative action plans have problems and shortcomings, and some analysts feel that they have all failed partly because they have not received adequate or continued federal support and partly because sanctions for noncompliance with mandatory standards were seldom involved (Foster, 1978). However, the more detailed examination of individual plans shows some gains in some cities, although the results are most often mixed and always below expectations and goals. The Washington, D.C. "imposed" plan was relatively successful in placing black workers even in the electromechanical construction trades, but since these workers do not have a formal union affiliation, the unions will have the last word about the success of this program (Rowan and Brudno, 1972).

The Indianapolis voluntary plan indicated the difficulty of retaining minority workers on the job once placed and the fact that it took threats at the federal level to motivate compliance of specific minority goals in several construction goals (Rowan and Brudno, 1972). The experience with the hometown plans in Detroit, Pittsburgh, and New York has been quite disappointing, with minority placements often reaching only a small percentage of the goals and an even smaller percentage of union memberships (Gould, 1977). Probably one of the best examples is the Boston hometown plan, a part of a proposed Model City program, which managed, despite the nonavailability of promised federal funds, to have the specified number of minorities enter the building trades as advanced trainees, trainees, and apprentices (Gould, 1977).

Finally in the 1970s, minorities seemed to have won a series of federal court decrees against the building trades, which have helped clarify the direct and indirect aspects of discrimination that lead to the exclusion of minorities. These legal trends not only are helpful in establishing legal principles but also can help eradicate discrimination through the potential legal threat that such discrimination may pose to unions and employers, provided the worst offenders are consistently identified and sued by the EEOC (Foster, 1978).

10.2.2 MINORITY CONTRACTORS: ARE WOMEN INCLUDED OR NOT INCLUDED?

The definition of the socially disadvantaged individuals as those "subjected to racial or ethnic prejudice or cultural bias because of their identity as a member of a group without regard to individual qualities" (New Minority Contracting Rules Won't Hurt Construction, 1978) in determining the new federal contracting rules could be equally applied to women as to blacks, chicanos, or American Indians. The federal rules, however, according to which 10% of public work contracts and other large federal contracts should be subcontracted to minority business, exclude women. The federal definition of minority excludes women and it specifies that 51% of the ownership and the active daily management of a business must be in minority hands (New Minority Contracting Rules Won't Hurt Construction, 1978). However, at least two states, California and Ohio, have proposed definitions of "socially disadvantaged" persons which include women as well as minorities (Gizyn, 1974; Minority Set-Asides, 1978). This proposed and enacted legislation is still too new to assess in any way its impact on minority (and women) contractors in construction.

10.2.3   WOMEN AND MINORITIES: WHO HAS BENEFITED MORE BY AFFIRMATIVE ACTION AND WHY?

The considerable head start of the affirmative action for minorities in the construction industry over that of women has led, in fact, to more tangible and substantial gains for minorities and at a much earlier date than has been true for women. Pressures to widen the access to minorities began in 1969 whereas similar pressures for women did not seriously begin until the late 1970s. For example, apprenticeships opened up for minorities between 1968 and 1971 even in the resistant electromechanical trades, in which the percentage changes ranged between 36 and 72% (Gruenberg, 1971). Such changes are still to happen for women.

Why does this difference exist, which cannot be attributed solely to the earlier beginning of affirmative action? Affirmative action for both minorities and women creates the same basic fears in employers, namely, that it will lead to the hiring of less qualified people with poor work performance and ability so that work productivity will suffer. However, because "minorities" suggests men to most people, there are some crucial differences in the way policies and programs are legitimized and implemented and in the way employers and co-workers react toward minority men versus women. Affirmative action plans and programs for minorities have been *relatively* more clearly defined and more rigorously implemented. Shorter training programs to supplement apprenticeships were early devised and successfully implemented (Lipsky and Rose, 1971), and some minority men were also informally trained to be journeymen and occasionally accepted as union members with the title of "trainee" or "advanced trainee." Women have had great difficulty establishing shorter training programs than apprenticeships and even greater difficulty getting informal, on-the-job training simply because they seldom were "sponsored" by skilled journeymen.

Furthermore, women have lacked the advantage of black men (and other minorities) of being hired by the nonunion residential construction industry; even at present, federal rules do not include women contractors among the socially disadvantaged contractors eligible for special contracting and subcontracting treatment.

It seems that the biological stereotypes are still very important in determining how affirmative action is implemented and what its outcome is. Minority men were always viewed as physically able to do the work whereas in the minds of most people there are still serious doubts whether women can do "men's work." Also, the stereotype that women do not support their families must still operate in shaping policies such as the exclusion of women contractors from the socially disadvantaged contractors. Finally, at the psychological and interpersonal level, the presence of women in a previously all-male occupational setting is much more disturbing than the presence of minority men. The main reason for this disturbance is the sexual tension that they create, a tension that men have to learn to manage in order to be able to work with them. Women's presence is also disturbing because men were socialized to protect and help women and have trouble learning to compete with them or to rely on them for help. Thus women do not get "sponsored" by journeymen because of the possibility of implied sexuality, or they get "sponsored" in exchange for sexuality. Women apprentices and women construction workers, on the other hand, have "problems" at work because they must often fight sexual harassment, and they must withstand open hostility, sabotage of their work, and lack of helpful co-workers. There can be no great gains for women in construction before these often subtle interpersonal patterns of discrimination are brought out into the open and dealt with. Otherwise, women's work performance will be criticized and women will not be accepted when in fact many of these problems are men's coping problems, which can be solved so that work productivity remains high.

ACKNOWLEDGMENT

The author wishes to acknowledge the assistance of Barbara O'Connor, who helped with the bibliographical search and with disentangling issues and data.

REFERENCES

A Record Apprenticeship Crop. *Engineering News Record*, 68, October 26, 1978.

AGC Serves Up Training Plans for Females and Minorities. *Engineering News Record*, 95–96, October 19, 1978.

Boserup, E. *Traditional Division of Work Between the Sexes, A Source of Inequality.* Geneva: Institute of Labour Studies (ILO), WDM/D2,FED/D2, 1976.

Boulding, E. et al. *Handbook of International Data on Women.* New York: Sage Publications, 1976.

Construction: The Industry and the Labor Force. A Reprint from the 1976 Employment and Training Report to the President. Washington, D.C.: U.S. Department of Labor, 1976.

Dubinsky, Irwin. Trade Union Discrimination in the Pittsburgh Construction Industry: How and Why It Operates. *Urban Affairs Quarterly*, **6**(3), 297–318, March 1971.

Ducharne, R. E. Problem Tools for Women. *Industrial Engineering*, **7**, 46–50, September 1975.

Equal Employment Opportunity in Apprenticeship and Training. *Federal Register*, Part IV, Friday, May 12, 1978. Washington, D.C.: Department of Labor, Office of the Secretary.

Foster, H. G. Industrial Relations in Construction: 1970–1977. *Industrial Relations*, **17**(1), 1–17, February 1978.

Foster, H. G. *Manpower in Homebuilding: A Preliminary Analysis.* Philadelphia: University of Pennsylvania Press, 1974.

Foster, H. G. and Strauss, G. Labor Problems in Construction: A Review. *Industrial Relations*, **11**(3), 289–313, 1972.

Gizyn, C. Women in Contracting: Growing Source of Management Skill. *DE/Journal*, **222**, 15–18, November 1973.

Gizyn, C. Women in Contracting, Part II. *DE/Journal*, **223**, 15–16, January 1974.

Gizyn, C. Women in Contracting, Part III. *DE/Journal*, **227**, 42, 46, February 1976.

Glover, R. W. and Marshall, R. The Response of Unions in the Construction Industry to Antidiscrimination Efforts. In L. J. Hausman et al. (Eds.), *Equal Rights and Industrial Relations*. Madison, Wis.: International Relations Research Association, 1977, pp. 121–140.

Gould, W. B. *Black Workers in White Unions.* Ithaca, N.Y.: Cornell University Press, 1977.

Gruenberg, G. W. Minority Training and Hiring in the Construction Industry. *Labor Law Journal*, **22**(8), 522–536, August, 1971.

Hammerman, H. Minority Workers in Construction Referral Unions. *Monthly Labor Review*, **95**(5), 17–26, May, 1972.

Hedges, J. N. and Bernis, S. E. Sex Stereotyping: Its Decline In Skilled Trades. *Monthly Labor Review*, **97**(5), 14–22, 1974.

Jackson, J. H. and Fossum, J. H. Attitudes of Apprenticeship Committee Members toward Affirmative Action Programs: A Preliminary Examination. *Labor Law Journal*, **27**(2), 84–88, February, 1976.

Labor Department Holds to Rigid Rules for Construction Hiring Plans. *Engineering News Record*, 7, February 9, 1978.

Labor Department Kills Two Hometown Plans, *Engineering News Record*, 212, September 21, 1978.

Landon, J. and Peirce, W. Discrimination, Monopsony and Union Power in The Building Trades. *Monthly Labor Review*, **95**(4), 24–26, April 1972.

Lipsky, D. B. and Rose, J. B. Craft Entry for Minorities: The Case of Project Justice. *Industrial Relations*, **10**(3), 327–337, October 1971.

Little Blond Widow Runs Thriving Construction Firm. *Engineering News Record*, 73–74, April 13, 1978.

Marshall, R., Glover, R. W., and Franklin, W. S. *Training and Entry into Union Construction* (Manpower R and D Monogr. 39). Washington, D.C.: U.S. Department of Labor, 1975.

Minorities and Women in Referral Units in Building Trade Unions. 1972 (Res. Rep. No. 44). Washington, D.C.: U.S. Equal Employment Opportunity Commission, June 1974.

Minority Set Asides: Ohio Legislature Holds Off on Minority-Only Contracts. *Engineering News-Record*, 64, July 20, 1978.

New Minority Contracting Rules Won't Hurt Construction. *Engineering News-Record*, 64, November 9, 1978.

Rich, L., Hardhatted Women in Construction. *Worklife*, **3**, 15–20, February 1978.

Riemer, J. W. "Deviance" as Fun—A Case of Building Construction Workers at Work. In K. Heney (Ed.), *Social Problems—Institutional and Interpersonal Perspectives*. Glenview, Ill. Scott and Foresman, 322–332, 1978.

Rowan, R. L., and Brudno, R. J. Fair Employment in Building: Imposed and Hometown Plans. *Industrial Relations*, **11**, 394–406, October 1972.

Safilios-Rothschild, C. *Women and Social Policy*. Englewood Cliffs, N.J.: Prentice-Hall, 1974.

Safilios-Rothschild, C. Young Women and Men aboard the U.S. Coast Guard Barque "Eagle". *Youth and Society*, **10**, 2, 191–204, 1978.

Shulins, N. Live-in-School Women Training in Building Trade, *Los Angeles Times*, Part I, p. 25, July 29, 1979.

State and National Apprenticeship Program System (SNAPS) Report. Washington, D.C.: U.S. Department of Labor, Employment and Training Administration, January 1–June 30, 1975.

The Changing and Contemporary Role of Women in African Development (mimeograph). UNECA, New York: 1974.

The Role of Women in African Development. ECA, Social Development section, Women's Programme Unit, Adis Abebba: 1975.

Unisex MBE Goal Hit, *Engineering News Record*, 55, June 1, 1978.

Women in Apprenticeship—Why Not? (Manpower Res. Monogr. No. 33). Washington, D.C.: U.S. Department of Labor, 1974.

Women Training in Building Trade. *Los Angeles Times*, Part I, p. 25, July 29, 1979.

INDEX